CREATIVE MARGINS

Cultural Production in Canadian Suburbs

Suburbs can be incubators of creativity: innovative and complex, but all too often underappreciated. In *Creative Margins*, Alison L. Bain documents the unique role of Canadian artists and cultural workers in suburban place-formation and dismantles mischaracterizations of suburbs as cultural wastelands.

Creative Margins interweaves stories of the challenges and opportunities presented by the creation of culture in suburbs, focusing on Etobicoke and Mississauga outside Toronto, and Surrey and North Vancouver outside Vancouver. The book investigates whether the creative process unfolds differently for suburban and urban cultural workers, as well as how this process is affected by the presence or absence of cultural infrastructure and planning initiatives.

Bain shows how suburban culture can enhance a city-region's vitality and sustainability. This book firmly debunks the myth of culture as a solely urban phenomenon and demonstrates the social and economic merits of investing in suburban art and culture.

ALISON L. BAIN is an associate professor in the Department of Geography at York University.

ALISON L. BAIN

Creative Margins

Cultural Production in Canadian Suburbs

UNIVERSITY OF TORONTO PRESS
Toronto Buffalo London

© University of Toronto Press 2013
Toronto Buffalo London
www.utppublishing.com
Printed in Canada

ISBN 978-1-4426-4690-2 (cloth)
ISBN 978-1-4426-1469-7 (paper)

∞

Printed on acid-free, 100% post-consumer recycled paper with
vegetable-based inks.

Library and Archives Canada Cataloguing in Publication

Bain, Alison L., 1974–, author
Creative margins : cultural production in Canadian suburbs / Alison L. Bain.

Includes bibliographical references and index.
ISBN 978-1-4426-4690-2 (bound). – ISBN 978-1-4426-1469-7 (pbk.)

1. Arts and society – Canada – Case studies. 2. Popular culture – Canada – Case
studies. 3. Artists – Canada – Case studies. 4. Arts – Economic aspects – Canada –
Case studies. 5. Suburbs – Canada – Case studies. 6. Canada – Cultural policy –
Case studies. 7. Sociology, Urban – Canada – Case studies. I. Title.

NX180.S6B34 2013   700.1′030971   C2013-903890-6

This book has been published with the help of a grant from the Canadian Federation
for the Humanities and Social Sciences, through the Awards to Scholarly Publications
Program, using funds provided by the Social Sciences and Humanities Research
Council of Canada.

University of Toronto Press acknowledges the financial assistance to its publishing
program of the Canada Council for the Arts and the Ontario
Arts Council.

University of Toronto Press acknowledges the financial support of the Government of
Canada through the Canada Book Fund for its publishing activities.

Contents

Illustrations

Figures

Table

Acknowledgments

What I have chosen to write about has come from the lives, experiences, and labours of others. I am grateful to the Social Sciences and Humanities Research Council of Canada for financial support and to the cultural workers from Etobicoke, Mississauga, North Vancouver, and Surrey who generously shared their stories and, in so doing, made this project possible.

Invaluable in the research process was my transcriber, Kelly Azoulay, and my graduate student research assistants: Tomas Jonsson conducted interviews in Mississauga, Meghan Rees coded transcripts, and William Sturm designed maps. Through presentations and discussions at the City Institute at York University, the Departments of Geography at York University and the University of Toronto, and the annual conferences of the AAG and the CAG, I learned from many people: Ranu Basu, Myrna Breitbart, Deb Cowen, Suzy Fitzpatrick, Harriet Hawkins, Brian Hracs, Doreen Jakob, Roger Keil, Deborah Leslie, David Ley, Norma Rantisi, Bas van Heur, and Douglas Young. For helpful conversations, I thank Margo Charlton, Lisa Choi, Rob Fiedler, Janet Friskney, Louise Johnson, Philip Kelly, Don Leffers, Lucia Lo, Libby Lunstrum, Dylann McLean, Heather McLean, Linda Peake, Valerie Preston, Steven Tufts, John Warkentin, and Tricia Wood.

I especially appreciate the time and effort that Jon Caufield put in to reading and commenting on the first draft of my manuscript, and the detailed and constructive recommendations made by three anonymous reviewers. I thank Lyndsay Wheelans for her photographs and for sharing the research journey with me. To Sarah and Douglas Bain, Ross Burnett, Allison Geuder, Nick Hanning, Julia Wakeham, and Carolyn Ziegler I extend special thanks for their ongoing friendship and support. It is to my parents, Anne and Tony Bain, for their years of unwavering love, guidance, and editorial advice that I dedicate this book.

CREATIVE MARGINS

Cultural Production in Canadian Suburbs

Chapter One

Introduction: Creating from the Margins

Let us think about margins as much as centres, of work as much as play, of ways of being in the city that do not correspond to bourgeois forms of entertainment – and let us find ways of imaging the city in these terms as well.

—(Williams, 2004, 241)

"Canada is a suburban nation" (Gordon & Janzen, 2011).[1] This increasingly common refrain, taken up by academics and journalists alike, is supported by scholarly research and census data that show that Canadian suburbs outnumber and outweigh cities: "More people live in the postwar suburbs than in the cities around which they grew" (Harris, 2007, A17). Census Canada demographic data show that between 2001 and 2006, nearly every Canadian suburb experienced a population boom. A similar trend has been underway for decades in the United States and Australia,[2] where "the dominant geography continues to be suburbia" (Kotkin, 2010a) and the population is said to be decentralizing faster than at any other time in human history (Flew, 2011).[3] In Canada, over half a century of intensive suburban expansion has dramatically changed the dynamics of cities and the peri-metropolitan areas that surround them (Bourne et al., 2011).

Canadian city-regions shelter diverse forms of suburban development and multifaceted suburban landscapes that offer significant research opportunities to extend critical analysis of the social, cultural, and spatial complexities of everyday suburban life (Fiedler & Addie, 2008). The recent book *In-Between Infrastructure: Urban Connectivity in an Age of Vulnerability* (2011) challenges academic and popular (mis)understandings of Canadian suburbs as sterile and uninteresting places by revealing some of the pressing infrastructural challenges faced in the transitional zone between post–Second World War suburbs and the exurban fringe. "These in-between spaces assemble a wild and often unexplainable mix of uses untypical for either the inner city or the classic suburb, they present landscapes of extreme spatial and social segregation" (Young et al., 2011, 3). As these authors productively illustrate, Canadian suburbs are not monolithic, fixed categories of urban space; they are neither unified nor uniform. Like their urban counterparts, suburbs are heterogeneous, complex, contradictory, and multi-authored places eminently worthy of continued re-conceptualization.[4]

One way to rethink suburbia, as Jennifer Robinson (2006) has persuasively argued with respect to urban centres in her book *Ordinary Cities*, is to develop an analytical framework that does not favour the experiences of certain kinds of cities or parts of cities over others. In this book, I advocate for an approach to studying suburbs that does not see the diffused urbanity of the exurban fringe as less culturally advanced or dynamic than the central city. The dichotomy of city and suburb is a powerful framing device, yet it also inadequately conveys the subtleties and complexities of new and emergent social, cultural, and spatial relationships in city-regions and it limits policy horizons (Kirby, 2004). A significant body of Australian scholarship on the cultural complexity of suburbia reveals the irrelevancy of historically constructed binaries between cities and suburbs, and essentialist imaginaries of suburbs and suburban residents (Powell, 1993; Johnson, 1994; 2012; Burnley et al., 1997; Stevenson, 1999; 2003; Gibson & Brennan-Horley, 2006). Inspired by this research, particularly empirical studies that focus on arts and culture and creative industries in Australian suburbs (e.g., Gibson & Brennan-Horley, 2006; Brennan-Horley & Gibson, 2009; Collis et al., 2010; Felton et al., 2010; Gibson, 2010; 2011; Brennan-Horley, 2011; Flew, 2011; Podkalicka, 2011; Flew 2012; Flew et al., 2012; Gibson et al., 2012), this book extends the conversation about the suburbs as complex, creative, and innovative places to Canada.

This book uses locality-based ethnographic analysis of the everyday practice of cultural production in Canadian suburbs to challenge an ordering of urban space that produces the suburban as "sub-creative," and a denigrated spatial "Other" (Phelps, 2012, 260). Its intention is to reveal suburban cultural complexity, creativity, and innovation in places that have been too readily marginalized, dismissed, and characterized as featureless, homogeneous cultural wastelands. This book offers a portrait of suburban creativity that reveals the practical, hidden, temporary, and spontaneous dimensions of the everyday working lives of cultural workers in suburbia (Shorthose, 2004). Fundamentally, this book is about the unique and constitutive role of Canadian cultural workers in suburban place-formation; it is about the suburbs as places in a continual process of becoming. As incomplete places that are in the making and not yet fully formed, suburbs are malleable and open to transformation. Suburbs possess a unique capacity to follow a different development trajectory and to become somewhere and something different from the urban. It is in accessing these place potentialities that the value of suburbs as a cultural resource can be realized. The chapters tell interwoven stories about the distinctive contributions of suburban cultural workers as place-makers who possess a keen sense of place, and they reveal some of the challenges and opportunities suburban cultural workers encounter through their creative practice in, and their artistic reconstitution of, Canadian suburbs. Taking inspiration from Edward Relph (1976; 1997), I understand sense of place as an innate, synthetic faculty for careful and critical observation of, and critical reflection on, the interrelated, changing, and embodied elements of places.[5]

It is easy to forget marginal spaces and the places contained within them, and through this forgetfulness to reinforce negative depictions that perpetuate neglect. While interest in the creative capacities and cultural economies of cities has surged since the publication of Richard Florida's (2002) oft-quoted and -critiqued book *The Rise of the Creative Class*, and urban scholars are increasingly interested in how the work of artists and cultural practitioners can help to develop new engagements with urban space and place, until recently, this scholarly literature has largely ignored the suburban experience of cultural production and consumption. While questions about where cultural workers choose "to live within and between cities, and how they relate to their communities" have been acknowledged as significant by the cultural economist Ann Markusen (2006, 1925), there remains insufficient additional scholarly work in North America that examines how the lived experiences of suburban cultural spaces help to anchor artists' creative development. In the United States, the McKnight Foundation (2002, 7) in Minneapolis, Minnesota asserts that "slowly and subtly, the arts have helped create a sense of place and build connections within more and more suburban communities. But the arts' importance to enhance the liveability of suburban communities hasn't been acknowledged as openly as it deserves." In an effort to speak to some of this neglect, I explore here the existence and potential of the alternative, spatially dispersed geography of cultural production in suburbia. Like Norcliffe & Rendace (2003), I define cultural production as acts of artistic creativity that combine intellectual and bodily labour to create goods and services with aesthetic appeal.[6]

Of Margins, Centres, and the Spaces Between

In the international world of cultural production, it is global cities in the northern hemisphere and the industrialized West, such as Paris, Berlin, Rome, London, and New York City, that have been renowned throughout history as centres of art and culture. Peter Hall (1998, 286) describes these creative cities of high culture as "places of great social and intellectual turbulence, not comfortable places at all." Yet Hall seems not to recognize that the experience of discomfort and challenge can also increase for individuals and social groups the greater their distance from the centre of a city, art world or global urban system. The dynamic intricacies of suburban art scenes in second-tier cultural centres have been pushed into the shadowy recesses beyond the glare of the global limelight and are rarely independently recognized as creative milieux. However, a growing body of contemporary academic work on creativity has sought to rectify this partial, metro-centric perspective by exploring how culture is produced in remote, rural, small, and suburban places (e.g., Gibson, 2002; Markusen & King, 2003; Gibson & Connell, 2004; Bell & Jayne, 2006 and 2010; Gibson & Brennan-Horley, 2006; Danaher, 2007; Brennan-Horley & Gibson, 2009; Luckman et al., 2009; Sorensen, 2009; Waitt & Gibson, 2009; Bennett, 2010; Mayes, 2010; Gibson, 2011; Bain & Marsh, 2012; Bain & McLean, 2012; Denis-Jacob, 2012; van Heur & Lorentzen, 2012). Leading the way is a special issue of the *Australian Geographer*, guest edited by Chris Gibson (2010, 3), that explores "the creative industries outside of major cities – in places that are physically and/or metaphorically remote, are small in population terms, or which because of socio-economic status or inherited industrial legacies are assumed by others to be unsophisticated, or marginal in an imaginary geography of creativity." This special issue was later published as the book *Creativity in Peripheral Places: Redefining the Creative Industries* (Gibson, 2011). Several of the Australian contributors to the special issue and book also participated in a themed issue of the *International Journal of Cultural Studies* entitled "Creative Suburbia: Cultural Research and Suburban Geographies" that examines the role of the creative industries in contemporary suburban development (Flew et al., 2012).

For cultural workers living in first-, second-, and third-tier cities in the global urban system, it is a growing challenge to obtain affordable and spatially central production and support space (e.g., studio, rehearsal, storage). Building upgrades, rent increases, condominium construction, neighbourhood gentrification, and municipal image makeovers through flagship cultural facility projects have all directly or indirectly contrib uted to the displacement and outmigration of cultural workers and arts organizations from central cities.[7] When the cultural sector labour force is required to relocate, individuals and organizations may look to neighbouring municipalities and suburban edge cities. Relying on a Richard Florida–inspired creative cities urban policy script, many Canadian suburban municipalities have responded with attempts to mobilize professional artistic activity as an economic driver, using copy-and-paste policy strategies that "'accessorize' neoliberal urbanism in a manner befitting cultural tropes of competitive cosmopolitanism" (Peck, 2011b, 62).[8] But in an era of "fast policy," where quick and easy-to-implement solutions to economic problems are favoured, entrepreneurial branding schemes often do little to substantially challenge the cultural authority of the central city (Oakley, 2009). The branding initiatives remain too reliant upon a limited palette of economic growth-oriented, mainstream creative activities that do little to reconfigure cultural hierarchies within city-regions. In this hegemonic cultural system, where the central city is celebrated as the legitimate locus of cultural production, the challenge for suburban municipalities is to resist the seductive ease of best-practice examples that are packaged by policy makers as essential to the maintenance of urban competitiveness.

In a chapter on "contradictory space" in *The Production of Space*, the French philosopher Henri Lefebvre (1991, 332) argues that while "the centre continues effectively to concentrate wealth, means of action, knowledge, information, and 'culture,'" it is important to recognize that the history of modern cities has shown that "centrality is movable." He goes on to explain that "throughout history centralities have always eventually disappeared – some displaced, some exploded, some subverted. They have perished sometimes on account of their excesses – through 'saturation' – and sometimes on account of their shortcomings" (Lefebvre, 1991, 333). Such "shifts in centrality" demonstrate, Lefebvre (1991, 332) asserts, a complex interplay between centre and periphery in which analytical categories remain unfixed and open to challenge.

On the edges of Canadian city-regions, logics of centrality have been disrupted by the dramatic material and demographic growth of suburban municipalities and the political and economic engines that support them. In advanced economies, underpinned by flows of information and investment capital, centres do not have to be attached to fixed points – they can be anywhere. Consequently, new paths and nodes, densities and interactions have begun to reconfigure social, cultural, and spatial relationships between the central city and the suburban periphery. The concept of periphery, like the concept of margin, is evocative, Rahamimoff (2005, 69) notes, of words like edge, frontier, fringe, "utmost limit, extremity – and a feeling of something risky, alert, yet also vulnerable." In the context of urban development, peripheries are "always moving, out-of-focus, volatile, and incomprehensible at a glance" (Ingersol, 2006, 9). Such changeability is valuable because it marks urban peripheries not just as end points but also as starting points.

In postmodern and postcolonial social theory, it is the margins, characterized by differences and distances from the mainstream, that are frequently celebrated as important locations of resistance and intervention in hegemonic narratives (Mitchell, 1997). Margins are vaunted for affording "critical 'locations' from which the relatively powerless can challenge dominant knowledge" (Hubbard et al., 2002, 78). Certainly as part of a liberal politics of difference, important scholarly and activist work has been done to bring people and places "at the margins back into cultural and political discussion, to put an end to the denial of the other's difference" (McGuinness, 2000, 226). Although margins have been interpreted as "crucial" for the "destructive and creative possibilities" they can realize (Sibley, 2001, 242), scholars have also warned against a "misplaced nostalgia for, or romanticisation of, the outside as a privileged site of radicality" (Fuss as cited in Binnie, 1997, 226). One cannot just assume the transgressiveness of the position of the outsider and the Other. There are shifting constellations of power and complex relationalities between insides and outsides, centres and margins that need to be negotiated, if people and places are to be thought about differently.

In their book *Postcapitalist Politics*, J.K. Gibson-Graham (2006, xxxvi) offer an important feminist critique of neoliberal global capitalism and its attendant power dynamics, hierarchies, and erasures. In collaboration, they have practised their politics by creating "spaces of collective endeavor and subjectivity, linking them to the project of constructing new economies." Gibson-Graham's (2006, 79) theorizations of the diverse "community economy" reveal hidden, yet productive and counter-hegemonic, dimensions of the economy in the spaces of everyday life. A discourse of the community economy, they argue, "works with a very different conception of the economy as an open system populated by diverse and incommensurable activities in which path-dependent relationships rather than predictable logics contribute to economic dynamism" (Gibson-Graham, 2006, 93). Gibson-Graham (2006, 79 and 88) affirm the great value of "resocializing economic relations" in ways that strategically negotiate plurality, interdependence, and the "ethics of connection." Inspired by their theorization of non-capitalist forms of exchange rooted in generosity and conviviality, this book documents a hybrid of commercial, domestic, and community economies that work to sustain creative practice and cultural production in suburbia. The intent of this book is to understand creativity as "more than just its market value, or only possible in contexts conducive to its marketability" and to appreciate "alternative spaces and logics for creative practice" (Danaher, 2007, 19). Following Gibson-Graham's (2006, xxxiii) understanding of place as "the site of becoming, the opening for politics," this book treats the suburbs as places of collective possibility with the potential to nurture diverse creative capacities and critical engagements with the world (Bromberg, 2010). In the contemporary period of spectacular urbanism, however, suburbs and the cultural work that is produced there continue to be relegated to the urban policy and investment shadows.

The Optics of the Competitive City

In the twenty-first century, cities are run like businesses. Local governments rely on the qualities of risk-taking, inventiveness, self-promotion, and profit-maximization and employ flagship development strategies centred on profit, spectacle, and consumption (Logon and Molotch, 1987; Kipfer and Keil, 2002; Hall, 2003). Around the world it is commonplace to use architectural form and urban design to promote a positive and a high-quality image of a city (Harvey, 1989; Zukin, 1995; Gibson, 2004). Urban planners frequently rely upon the expert advice of architects, who are prone to recommend beauty, design excellence, and architectural drama as a discursive strategy to promote growth and to mute opposition to it (Boudreau et al., 2009). This attention to place-making treats the city as an aesthetic branding project where a distinctive civic image is developed to appeal to footloose tourists, investors, and members of the creative class (Evans, 2007). The goal of such place-branding exercises is usually to replace perceptions of the city as a place of disinvestment, decay, crime, and poverty with images of growth, vitality, and prosperity (Avraham, 2002). It is not just the built and the physical environment that is central to imaging a city. The lifestyles and symbolic economies that are consumed in these spaces through recreation, leisure, and cultural activities also contribute to a city's image and the perceived quality of life (Zukin, 1995; 2004). Increasingly, as Cronin and Hetherington (2008, x) document, "forms of spectacle have come to shape how cities are imagined and to influence their character and the practices through which we know them." Spectacle now dominates the urban landscape.

In *The Society of the Spectacle*, Guy Debord (1994) argues that spectacle is both the dominant characteristic and the primary product of contemporary society. For Debord (1994), spectacle is a glittering distraction of "false choices" and banal trends that deceitfully inverts reality with its illusions and images, blankets social spaces with layers of commodities, and alienates people with a duty to consume. Debord's concept of spectacle is fundamentally connected, Knox (2008, 102) explains, to notions of "separation and passivity, for in submissively consuming spectacles, one is estranged from actively producing one's life" and is absorbed with appearances rather than political engagement.

Culture, in particular, has become an instrument of economic development and "urban spectacularisation, which serves both for real-estate speculation and for political propaganda" (Vaz & Jacques, 2007, 249). For over two decades, academics, urban planners, and municipal politicians around the world have embraced a flexible creative capital model of urban development that privileges knowledge, creativity, and commodified difference as a means of civic renewal (Gertler et al., 2002; Markusen & King, 2003). This model turns away from the more traditional primary and secondary sectors of the economy for support and seeks instead to attract a broad "creative class" of professionals in business, law, engineering, science, health care, architecture, design, and the arts who supposedly share similar approaches to complex problemsolving and a work ethos that values individuality, difference, and merit (Florida 2002; 2005). The iconic citizens of the new knowledge-based economy are presented as the "storm-troopers" of city branding; they are said to be attracted to cities that offer the "3Ts" (technology, talent, and tolerance). But as Wilson and Keil (2008) rightly question, tolerance by whom and of what? For these scholars, the "real" creative class is the poor – who through everyday resourcefulness and ingenuity negotiate torn social safety nets, navigate neglected neighbourhoods, and labour in low-wage service sector jobs. Wilson and Keil (2008, 846) argue that the concept of the creative class is a perverse fantasy in which "the white affluent are the historic and current bearers of civility, tradition, and good culture"; and the creative capital model of urban development is just another fast-track public policy option that privileges "the desires and aspirations of capital and the affluent."

In a quest to lure the young, well-educated, hipster workers of the knowledge economy, many large Canadian cities are remaking themselves as centres of arts and culture, marketing themselves as places that provide stimulation, diversity, and a richness of experience that inspires creativity and innovation (Broudeau et al., 2009). Nowhere is this more boastfully displayed in Canada than in Toronto.

Cultural Boosterism in Toronto

The cultural turn in Toronto's urban economic development strategy began in the 1970s when the city spent money on the arts and sought to connect culture to a night-time economy. Traditional high-cultural forms (e.g., theatre, opera, dance, literature, and the visual arts) flourished and spaces for experimental and interactive arts proliferated (e.g., comedy clubs, fringe theatre, repertory cinemas, and night clubs). The annual Toronto International Film Festival (TIFF), for example, was established in 1976. By attracting funding agencies and corporate sponsors with the promise of spectacle and large financial returns, TIFF has since become the largest film festival in North America. It now has a permanent home in the Bell Lightbox, at the base of a mixed-use condominium Festival Tower (Mays, 2009, G2).[9]

Through the 1980s, Toronto's reputation as a film, media, and theatre centre gained currency. By the end of that decade it was popularly claimed that Toronto was "Broadway North" – the third largest theatre centre in the English-speaking world, after London and New York (McKinnie, 2007). The historic Elgin and Canon (formerly Pantages) theatres were restored, the Mirvish family built the Princess of Wales Theatre on King Street West in the rapidly growing Entertainment District, and performing arts centres were built in the suburbs of North York and Mississauga. Through the neoliberal restructuring of the 1990s, construction of large cultural projects slowed and funding for the arts was substantially cut back. During the "Common Sense Revolution" of Ontario's Progressive Conservative Premier Mike Harris the arts were treated as a luxury (Milroy, 2009).

To initiate a dialogue around the potential of culture to transform urban economics and cityscapes, Toronto's city council commissioned a Culture Plan in 2000, which it adopted in 2003. The 10-year plan details recommendations for repositioning the city as an international cultural capital through the leveraging of resources, cultural facilities, public-private partnerships, financing, and heritage preservation. A component of Toronto's cultural plan involves the development of a Museum Arts District and an increase in funding to "Major Cultural Organizations" (City of Toronto Culture Division, 2005, 6).[10] In conjunction with the city's investment in and adoption of a cultural plan, in 2002 the provincial and federal governments jointly announced $233 million in financial support for seven cultural regeneration projects. With this injection of funding, Toronto witnessed an urban cultural building boom – what Stanwick and Flores (2007) describe as a "design renaissance" that seeks to socially and architecturally give cultural spaces new prominence in the city.

High-culture institutions in Toronto (e.g., the Royal Ontario Museum, the Art Gallery of Ontario,[11] the Gardiner Museum of Ceramic Art, the Four Seasons Centre for the Performing Arts, the National Ballet School, and the Royal Conservatory of Music) have undergone "starchitectural" makeovers by some of the world's brand-name architectural practitioners including Frank Gehry, Daniel Libeskind, Lord Norman Foster, and Will Alsop. Inviting Daniel Libeskind to write the forward to their book *Design City: Toronto*, Stanwick and Flores (2007: 8) build on his optimism about "Toronto the Great" when they write: "Toronto is now knee-deep in a phenomenal act of city (re)making. If a renaissance is about a rebirth and enlightenment, then Toronto's architectural renaissance is certainly on its way to maturity." The ambitious flamboyance of this quotation conveys a sense of the relentless enthusiasm of civic boosters for repositioning Toronto on the world stage.

This trend of employing high-profile international architects to create iconic architectural landmarks (what Roodhouse [2009] has termed "cultural iconographic regeneration") is not new. Architecture as urban change agent gained prominence with the successful completion in 1997 of the franchised Guggenheim Museum in the industrial city of Bilbao, Spain by the architect Frank Gehry (Vicario & Monje, 2003). The "Bilbao effect" uses the museum as a strategic marketing tool for a city; it "illustrates how a museum can come to represent a city the way cathedrals once had the power to signify" (Ingersol, 2006, 52). Former mayor of Winnipeg turned urban strategist Glen Murray has humorously referred to the mistaken belief held by civic enthusiasts that constructing "cultural trophy buildings" (Wright, 2008, 130) can be a shortcut to urban revitalization as the "Irritable Bilbao Syndrome." Instead of developing cultural planning strategies that use grassroots economic and cultural assets and resources for new uses, many urban decision makers in Canada's largest cities are busy with iconic interventions, elevating buildings of undulating metal, glass prisms, and colourful pillars as civic boosterism and branding projects.

Large-scale redevelopment initiatives that may inadequately attend to the specificities of local context have become central to how cities re-image themselves (Swyngedouw, 2007). Back in the 1970s, critics of cultural architecture advocated for a move away from "big, baronial, and inflexible palaces" in favour of "smaller, more intimate, less opulent and more flexible spaces" (Schafer, 1973 cited in Duxbury, 2008, 4); yet their voices went largely unheard and their criticism remains as valid today. Journalist Philip Kennicott (2009) echoes some of this criticism when he writes: "The 'starchitect' neologism that seemed to define the decade, also became something of a dirty word, as momentum grew for a new kind of modesty and problem-solving, rather than flamboyance and busted budgets." In Toronto, architectural modesty and problem-solving are in evidence in renovations to the elite cultural training facilities of Canada's National Ballet School and the Royal Conservatory of Music. However, naming rights and trends in private sponsorship reinforce a propensity to over-build cultural edifices for performance and presentation and, in the process, to neglect practical community needs for smaller preparatory workspaces for the arts. Again, consumption and spectacle are often mistakenly privileged over production and practice. A sustainable creative economy requires a shift in investment focus away from consumptive-oriented tourist spaces to the behind-the-scenes spaces of artistic preparation, creation, and production. The spaces where the creative labour force works after they have received their training should be the building blocks of the creative economy.

In Toronto, civic enthusiasm for and investment in arts and culture was strategically nurtured by Mayor David Miller over his seven years in office (2003–10). The "TO live with culture" campaign, Scotiabank Nuit Blanche, L'Oréal Luminato, and the Creative Spaces and Places conference are a handful of the signature cultural activities that the city supported in an effort to underscore its title as a Cultural Capital of Canada for 2005–6.[12] The promotion of culture and creativity as a new brand for the city through "fresh messaging and strong visuals" was one of the primary goals of Toronto's 2003 cultural plan (City of Toronto Culture Division, 2005, 5).

Initiated in the fall of 2005, the centrepiece of Toronto's cultural pro-motion package was the $4 million, sixteen-month street banner brand-ing campaign, "TO live with culture," and its accompanying web portal that listed arts events in the city (www.LiveWithCulture.ca). The objective of this marketing and promotion campaign was to increase awareness for locals and tourists alike of the extent of Toronto's cultural activities. In a satirical response to this high-profile event, Toronto-based writer and video artist RM Vaughan fought the commodification of culture with his own "Live without culture" counter-campaign. He photographed him-self in sweatpants as "an ugly, poor, underappreciated, misrepresented, hopeless loser" (Vaughan, 2006, 26), copied the fonts and colours from the official website, and produced his own spoof banners and buttons for sale. As Vaughan (2006, 25, 27) explains:

> Art is often very difficult and very, very boring, and culture in general requi-res a certain level of education ... Now everyone is part of the Creative Class. Culture is the new IT. It's a boom market. I bet some of the top peo-ple in the Creative Class actually look like the people on your banners: all healthy, full of fun and just bursting with creativity, at one with the muses.

The purpose of Vaughan's project was to provide a more realistic por-trait of the challenges, frustrations, and precariousness of artistic labour. He sought to highlight the shortcomings of a creative competitive pol-icy model that values only the commercial dimensions of culture for its marketability.

Integral to Toronto's culture branding campaign is a free, all-night, corporately sponsored celebration of contemporary art organized through the City of Toronto's Special Events Division. This "all-night con-temporary art thing" (www.scotiabanknuitblanche.ca) was called Nuit Blanche in its first incarnation in 2006, and in subsequent years rebranded Scotiabank Nuit Blanche. Paris first introduced the idea of an all-night art party in 2002 and the event has since been copied around the world in cities such as Madrid, Riga, Rome, Brussels, Montreal, and, now, Toronto. For one fall night, over a twelve-hour period from sun-set to sunrise, residents and visitors are invited to (re)discover Toronto through public art commissions, all-night exhibitions, performances, and installations located in designated zones in the central city. Widely regarded by politicians and sponsors as a success, the event continues to receive mainly laudatory coverage in the print media.

Voices of critical dissent within the arts community have, however, highlighted how the event legitimizes the "smooth, uninterrupted development of the corporate capital and city governance agendas in urban socio-economic reconfigurations/redevelopments (i.e., gentrification) and the whitewashing/erasing of meaningful/resistant difference and diversity (i.e., erasure of counter-culture and diversity through appropriation under city-corporate banners)" (Moradi, 2007, 7). Thus, while some isolated art works at Nuit Blanche have challenged political agendas, the majority have celebrated controlled spectacle and familiar interactions; artists are treated as attractions, audiences as consumers, and art as economic stimulus. This leads Adrian Blackwell (2006, 29) to question whether "the role of art [is] primarily to stimulate economic growth with deleterious social effects, or is culture to be considered first as that complex lattice of diverse practices required to produce healthy social spaces from which new economic opportunities will follow?" When employed first within a social rather than an economic framework, there is so much more that culture has the potential to accomplish in building community.

Short-term cultural events like L'Oréal Luminato and Scotiabank Nuit Blanche have certainly drawn larger crowds with each passing year and have temporarily increased local cultural awareness. But members of Toronto's arts communities argue that city funding would be better spent on capital grants that would allow arts organizations and cultural workers to break the cycle that displaces them from gentrifying neighbourhoods by purchasing permanent affordable workspace, earmarked for emerging artists (Barmak, 2008). Without such financial support, Toronto's street banners could potentially become a misnomer. In the future, culture and cultural workers may not be able to afford to live downtown.

Place-Making in Vancouver

In contemporary Canadian cities like Toronto and Vancouver, where significant inner-city redevelopment has occurred, the nineteenth-century middle-class stigma once associated with living among the supposed deprivation, pollution, and pathologies of the working classes has lost some of its potency. David Ley (1988, 1993, 1996a, 1996b) has documented how, since the 1970s, the romanticized cosmopolitan character of downtown living has gained cachet among the "new middle class," persuasively arguing that it is the residential preferences, investment decisions, and cultural values of this new middle class that has facilitated substantial gentrification and upgrading in many Canadian cities. Some of urban Canada's most expensive real estate is found in core-area neighbourhoods. Recognizing that land is a finite resource, many Canadian cities have focused development on brownfield and infill sites. Much of this intensification has taken the form of privately developed condominiums.

Condo-mania has a firm grip on both Toronto and Vancouver (White, 2008). The urban planning departments in both cities contend with condominium development as a major component of municipal comprehensive plans (Kern, 2005; 2007; 2010). The conversion of formerly productive industrial, manufacturing, and office use to high-density, middle- and upper-class residential use involves substantial physical change to the built and social fabric of cities.

Vancouver is an impressive example of place-making through condominium infill and waterfront redevelopment. This is a city that has been transformed from a control, distribution, and processing centre for natural resources in the staples economy to a "post-corporate city" of "mega-project redevelopment schemes" shaped by transnational flows of people, money, and ideas from the Pacific Rim (Barnes & Hutton, 2009, 1257). Over the last half-century, federal, provincial, and local governments have used narratives of globalization and transnationalism in conjunction with hallmark events to rebrand Vancouver as a globalizing "Asia-Pacific city and as Canada's Pacific gateway metropolis" (Hutton, 2012, 279).

Since the 1990s, Vancouver has gained an international reputation for its high standard of urban planning and urban design practice, so much so that when American architects and planners promote the idea of a high-residential density, high-public amenity central city they call it "Vancouverism" (Boddy, 2007). In Vancouver, the mountains and the ocean provide the raw material for the "cult of the view," which, in conjunction with a series of overlapping official "view corridors," has driven high-rise residential development in the central city for the last fifty years (Berelowitz, 2005). Historic neighbourhoods in the central city (e.g., Strathcona, Gastown, and Chinatown) have been gentrified, but remain largely intact as pedestrian- and bicycle-friendly places because Vancouver has managed to avoid some of the worst of traditional North American urban renewal (e.g., highways, elevated and underground pedestrian systems, large shopping malls, and big-box retail). Unlike in Toronto, in the City of Vancouver there are no highways within the municipal boundaries to disrupt the regular street grid or to cut the city off from its waterfront. Since the 1970s and accelerating after the Expo '86 World's Fair and into the 2010 Winter Olympics, Vancouver has strategically transformed its waterfront from industrial and railway use to residential and recreational use. The construction of public pathways along the seawall has repositioned recreational activity in a continuous corridor along the inner-city waterfront. The downtown street grid and urban fabric have been extended to the water's edge and filled in with tall, bluish-green glass condominium towers, known locally as "see-throughs" (Coupland, 2000). The city has faced criticism for privileging residential development over office development, thus jeopardizing its commercial land base, elevating land values, and creating more of a recreational resort than a working city.

Against a backdrop of yacht clubs, glass towers, and cloud-shrouded mountain peaks, a centrepiece of Vancouver's waterfront renewal is Granville Island. This former industrial site beneath the southern viaduct of the Granville Bridge was transformed in the 1980s into a commercial, arts, entertainment, and recreation destination. The foci of this complex are a large public farmers' market and the Emily Carr Institute of Art and Design, together with an eclectic mix of studios, galleries, theatres, restaurants, pubs, and specialty shops in converted warehouses. Granville Island has been characterized as a privatized public place of conspicuous consumption and play (Ley, 1996a). Facing Granville Island, on the north side of False Creek, is Concord Pacific Place. This former Canadian Pacific Railway yard and then 1986 World Exposition site was redeveloped by the Hong Kong multi-billionaire property developer Li-Ka Shing into high-rise condominium towers for "empty nesters, professional singles, and a new cohort of wealthy East Asian immigrants" (Ley, 2010, 17). Located on two kilometres of waterfront and stretching over 10 city blocks, Concord Pacific Place showcases the power and capital of transnational elites to redefine and reconstruct downtown Vancouver's built landscape (Berelowitz, 2005).

Concord Pacific Place abuts another neighbourhood, Yaletown, on the southeast side of the downtown peninsula, that typifies, Hutton (2004b, 1955) argues, "the rise of a specialized inner-city 'new economy' shaped by intersections between culture and technology." For Trevor Barnes and Thomas Hutton (2009, 1257), this new economy emphasizes "experimentation, innovation, creativity, applied design and the synthesis of production and consumption." In Yaletown, brick warehouses, industrial plants, and working-class housing have been variously refurbished or demolished and replaced by condominiums, creative industries firms (e.g., software and video game development, computer systems design, architecture, engineering, graphic design, and advertising), high-end restaurants, supermarkets, coffee shops, and fitness centres all catering to the consumptive needs of the creative class (Hutton, 2004a). Yaletown, Concord Pacific Place, and Granville Island are just a small sampling of territorial bases of wealth, investment, and pleasure in Vancouver that stand in marked contrast to the poverty, homelessness, and struggles with addiction experienced by many residents of the Downtown Eastside, perhaps Canada's poorest inner-city neighbourhood (H. Smith, 2003; Ley & Dobson, 2008).

On international indices of quality of life and "liveability," Vancouver continues to rank highly (Hutton, 2012). Nevertheless, it is also a city renowned for the high cost of its central-city real estate. As a result, it is usually only the very affluent who can afford to call central Vancouver home. In Vancouver, as in Toronto, the process of urban redevelopment has physically reorganized poverty and affluence in the city, often concentrating poor residents in neighbourhoods that are poorly serviced, far from employment opportunities, and physically separated by modernist planning practice from the rest of the city (Walks, 2001). The image of a new, affluent city of leisure seeks to hide such marginal spaces from view.

Redistributing the Gaze to Appreciate the Unspectacular

To withstand waves of national and international competitive urbanism, Toronto and Vancouver have relied on branding and spectacularization to aid in the reinvention and redevelopment of their urban fabric. In Toronto and Vancouver, city-building has emphasized architectural form and modes of theatrical display that dramatize the city, highlighting grand urban gestures over more practical forms of public life and dwelling (Williams, 2004). The dramatic visibility of cultural megaprojects, luxury condominiums, hotel and retail districts, and festival waterfronts entices political decision makers and private developers, who remain convinced of the economic and cultural dividends that such large-scale central-city investments can generate for municipal, regional, and national economies. The shiny and new often get municipal funding priority over the unsexy safety upgrades, retrofits, and deferred maintenance needs of older buildings, yet "their impact is mostly negligible on the daily lives of the vast majority of urbanites who live in the mid-to-outer-ring suburbs" (Fien, 2005, 236).

During the last four decades much urban planning, policy, and investment attention has been directed to the élite, high-value spaces in the "monumental core of cities" (Ingersol, 2006, 64) – what Christine Boyer (1995) has termed the "figured" city. For Boyer (1995, 81), the figured city is composed of "well-designed and self-enclosed places" that are "imageable and remembered" to the affluent and powerful people who live, work, and play within them. The disfigured city, on the other hand, is composed of "the interstitial spaces [that] are abandoned or neglected" which surround and penetrate the figured city (Boyer, 1995, 81). The disfigured city, like the suburbs that frame the central core, remain "unimageable and forgotten," "invisible and excluded" (Boyer, 1995, 81).

It is time, Richard Ingersol (2006, 64) provocatively reminds us in *Sprawltown: Looking for the City on Its Edges*, to "consider the inverse movements that shift the attention away from the centre." It is time to "de-marginalize the in-between city and reshape the perception of it as something other than a social and economic threat ... [that] remain[s] in the shadow of ... glamour zones" (Boudreau et al., 2009, 140). It is time to turn our practical and scholarly attention away from the repeated adulation of the central city in order to develop a more nuanced appreciation of the changes occurring, and the creative possibilities awaiting, in the suburban periphery.

Living on the Edge: Canadian Suburban Cultural Workers in Toronto and Vancouver

There is a lot more cultural work happening in suburbia than the North American urban scholarly literature reveals. This book takes as its subject the Canadian suburban cultural worker and examines the relationship between creative labour and the suburban geographical context. Suburban cultural workers fall into the category of what sociologists refer to as the "unmarked" subject. Without an exceptional quality that might attract study, the unmarked subject can go undocumented (Lang, 2003). Who or what is a cultural worker?

This book takes a pragmatic approach to the definition of the culturally creative labour force. Instead of engaging with philosophical debates in art history or aesthetics about the conceptual meaning of artist, I use the term cultural worker because it is deliberately inclusive in its disciplinary reach. It accentuates the importance of valuing a diversity of creative activities as work that demands mental and physical labour worthy of remuneration. Pierre-Michel Menger's (1999) study of artistic labour markets and careers has shown that cultural workers as an occupational group are younger and better educated than the general workforce. They "tend to be concentrated in a few metropolitan areas, show higher rates of self-employment, higher rates of unemployment, several forms of constrained underemployment, and are more often multiple job holders. They earn less than workers in their reference occupational category, that of professional, kindred, and technical workers, whose members have comparable human capital characteristics (education, training, and age), and have larger income inequality and variability" (Menger, 1999, 545). Menger's generalizations about the artistic labour force find support in statistical profiles of Canadian artists produced using Statistics Canada census data for the Canada Council for the Arts (Hill Strategies Research Inc., 2006; Hill & Capriotti, 2009).

A report titled "Artists in Large Canadian Cities" (Hill Strategies Research Inc., 2006) provides an overview of arts employment in 92 Canadian cities with populations over 50,000. The report determined, using nine professional arts occupational categories,[13] that in 2001 there were 130,700 artists in Canada, earning on average $23,500 (26 per cent below overall labour force earnings), and representing 0.8 per cent of the labour force.[14] Between 1991 and 2001 the number of artists in Canada grew 29 per cent, a rate of growth that trebled that which occurred in the overall labour force. A follow-up report based on the 2006 census showed that the number of people in Canada who spend more time at their art than any other occupation rose to 140,000, still representing 0.8 per cent of the Canadian labour force, and earning on average $22,700 (a 37 per cent earning gap with the rest of the labour force). Between 2001 and 2006 the number of artists in Canada increased by 7 per cent, which is slightly less than the 11 per cent increase experienced by the overall labour force (Hill & Capriotti, 2009).[15] Although studies indicate that many Canadian visual artists reside in rural locales (Bunting & Mitchell, 2001; Mitchell et al., 2004), the majority of Canadian cultural workers do live and work in one of the three largest census metropolitan areas (CMAs) – Toronto (35 per cent), Vancouver (16 per cent), or Montreal (27 per cent).

In absolute numbers, Toronto has Canada's largest concentration of cultural workers. According to Statistics Canada census data, the number of arts and culture professionals in the Toronto city-region grew from

42,810 in 2001 to 48,575 in 2006, an increase of 13 per cent, with over 15,500 cultural workers living outside the City of Toronto in surrounding suburban communities. In the Vancouver city-region, the number of arts and culture professionals had a similar percentage increase to that in Toronto, growing 13 per cent from 20,050 in 2001 to 22,755 in 2006, with over 11,000 cultural workers living outside the City of Vancouver in 2006. The four suburban case study sites discussed in detail in this book include an inner and an outer suburb from Toronto and Vancouver, where there are noteworthy concentrations of people employed in arts and culture professions. These suburbs were identified primarily based on their location in the city-region and the government boundaries that define them. A distinction is made between inner and outer suburbs based on the general age of the housing stock and built fabric and the relative distance from downtown (Bunting et al., 2004). Inner suburbs may have older historic roots as townships or municipalities in their own right, are located just outside of the central core of cities, and experienced significant growth in the period between 1940 and 1960. Beyond the inner ring of postwar suburbs, outer suburbs grew dramatically after 1970.

From Ontario, this book's Toronto suburban examples are Etobicoke and Mississauga, while from British Columbia, Vancouver's suburban examples are North Vancouver and Surrey. The inner suburb of Etobicoke on the western edge of the City of Toronto had 2,325 people professionally employed in arts and culture occupations in 2006 (1.4 per cent of the labour force), up 15 per cent from 2,025 (1.2 per cent of the labour force) cultural workers in 2001. The outer suburban municipality of Mississauga saw employment in professional arts and culture occupations increase 18 per cent from 2,735 people (0.8 per cent of the labour force) in 2001 to 3,230 people (0.9 per cent of the labour force) in 2006. The City and District of North Vancouver saw arts and culture employment increase 12 per cent from 1,600 people (2.2 per cent of the labour force) in 2001 to 1,790 people (2.5 per cent of the labour force) in 2006. In the outer suburban municipality of Surrey, arts and culture employment increased 16 per cent from 1,430 people (0.8 per cent of the labour force) in 2001 to 1,655 people (0.8 per cent of the labour force) in 2006. In both city-regions, it is the inner rather than outer suburbs that possess the greatest percentage of cultural workers in their labour force at percentages that are above the national average. Dot density maps in Figures 1.1, 1.2, 1.3, and 1.4 help to illustrate the spatial distribution of arts and cultural employment in the CMAs of Toronto

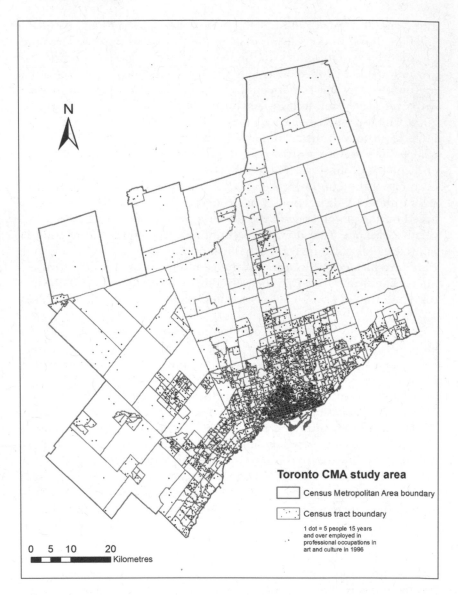

Figure 1.1   Dot density map of professional employment in art and culture in 1996 by census tract in the Toronto CMA (map by W. Sturm using Statistics Canada, 1996).

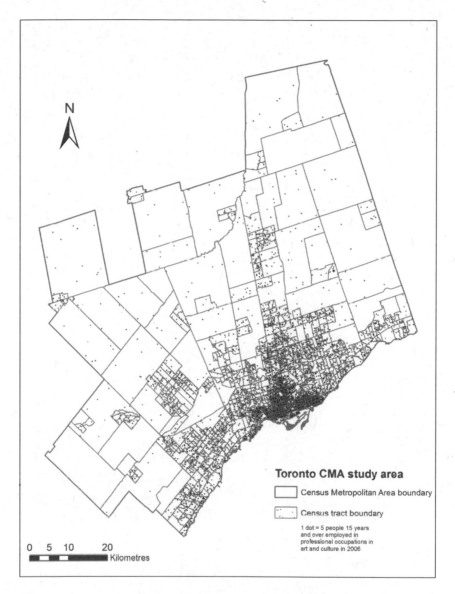

Figure 1.2 Dot density map of professional employment in art and culture in 2006 by census tract in the Toronto CMA (map by W. Sturm using Statistics Canada, 2006).

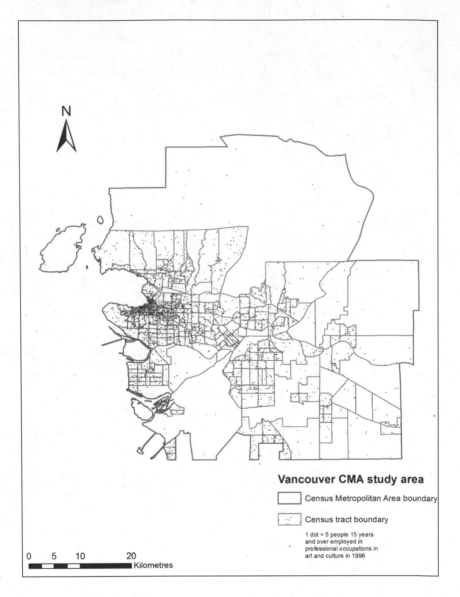

N

Vancouver CMA study area

Census Metropolitan Area boundary

Census tract boundary

1 dot = 5 people 15 years
and over employed in
professional occupations in
art and culture in 1996

0   5   10   20
▬▬▬▬▬▬▬▬▬ Kilometres

Figure 1.3   Dot density map of professional employment in art and culture in 1996 by census tract in the Vancouver CMA (map by W. Sturm using Statistics Canada, 1996).

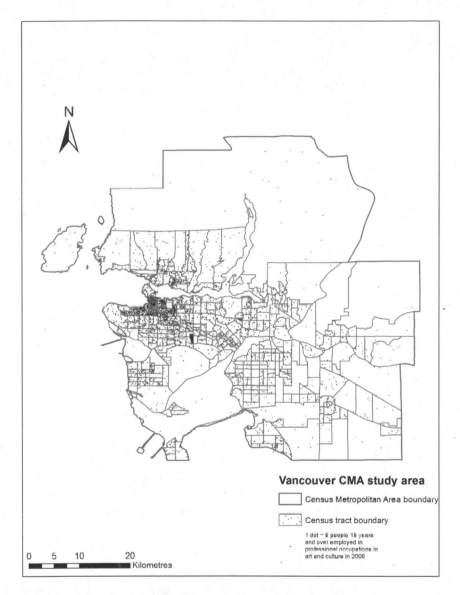

N

Vancouver CMA study area

☐ Census Metropolitan Area boundary

[⋅ ⋅] Census tract boundary

1 dot = 5 people 15 years
and over employed in
professional occupations in
art and culture in 2006

0   5   10   20
▬▬▬▬▬ Kilometres

Figure 1.4   Dot density map of professional employment in art and culture in
2006 by census tract in the Vancouver CMA (map by W. Sturm using Statistics
Canada, 2006).

and Vancouver in the decade between 1996 and 2006, showing a gradual increase in the numbers of suburban cultural workers.

In the city-regions of Toronto and Vancouver the cultural workers who were interviewed for this book all self-identify as artists and endeavour to practise the arts as a professional occupation. Art narrates the lives of these individuals and propels them forward in their professional practice. However, unlike other professions such as law or medicine that are regulated and certified by professional bodies, the arts do not require formal degree accreditation or an official demonstration of competence (Karttunen, 1998). While formal training opportunities may be on the increase in the arts, for many art forms training is advantageous but not a prerequisite for a sustained and successful career. Thus inclusion in this study did not require the demonstration of a connection to a training institution, participation in a certification process, or membership in a professional organization. Instead, the primary criterion for participation was self-evaluation of commitment to the arts both as a central life activity and as a publicly proclaimed profession, irrespective of whether it is the main source of income.

The study participants practise a range of different art forms including dance (ballet, modern, Indian classical), music (classical, jazz, rock), drama (acting for the stage, film, and television, storytelling), writing (poetry, fiction, non-fiction), fine arts (sculpting, printmaking, photography, painting, mixed media, installation, and drawing). Within the arts, the boundaries between different subdisciplines can be blurred. Some of the cultural workers work in more than one medium – combining storytelling and music, for example, or set design and mixed media fine arts. Regardless of the disciplinary combinations, each cultural worker identified a primary creative medium. Of the 57 cultural workers interviewed, 28 are visual artists, 8 are actors, 6 are musicians, 4 are writers, 4 are dancer/choreographers, 3 are potters, and the remaining 4 are each respectively a composer, conductor, filmmaker, and storyteller. In this sample of cultural workers, 60 per cent have a post-secondary education, 70 per cent are self-employed and do additional unpaid volunteer work, and nearly 50 per cent move in and out of part-time, casual, or short-term contract employment. Of the remaining 32 study participants interviewed, 28 are arts administrators (e.g., curators, directors of community arts organizations and arts councils, board members, and arts educators), and 4 are municipally employed cultural planners.[16] In total, 89 interviews were undertaken, 42 in Toronto (19 in Etobicoke and 23 in Mississauga) and 47 in Vancouver (26 in North Vancouver and 21 in

Surrey). These interviews provide the discursive foundation upon which this book is based.

Inspired by Paul Mattingly's (1997, 41) comment that "it is remarkable how many case studies of suburban communities have been completed without a serious engagement with the oral histories of suburbanites," this book positions at centre stage the narratives of suburban cultural workers. A selection of suburban cultural workers were invited to speak in detail and at length about the suburbs in which they live and work; they narrated the material, social, and policy landscapes that form the backdrops to their lives. Although suburban cultural workers often may not be the most affluent members of society, this book asserts that it is important to consider their relative power and influence in shaping the suburban landscapes of which they are an integral part. If, as Roger Silverstone (1997, 13) asserts, suburbia should be understood as "a state of mind" that is "constructed in imagination" and in the everyday lives of people, then cultural workers, whose primary creative resource is their imagination, have much to contribute to the suburban place-making exercise.

The Story

In the chapters that follow, a story unfolds about the relationship of culture and cultural workers to suburban places in Canada. To set the stage, chapter 2 provides an overview of relevant suburban studies literature and introduces the four main case study locations upon which this book is based. This second chapter documents how the classic Anglo-American scholarly accounts of the history of North American suburban development neglect to adequately consider culture as a transformational force. Before delving into the details of the four case studies, chapter 3 makes a Trans-Canada pit stop on the intellectual journey between Toronto and Vancouver in the Calgary outer suburb of Arbour Lake. Discussion of the suburban artistic interventions of a local art collective, along with reflections on the metaphors of exploration and frontier, are used to consider why artists are more widely represented as urban pioneers than suburban pioneers. This theme of artists using suburbia as a creative foil is picked up again in subsequent chapters that continue to examine how alternative spaces of cultural production can augment the complexity of suburbs.

At the heart of this book is a straightforward set of questions about the advantages and disadvantages of creating art in a suburban location.

To answer these fundamental locational questions, chapter 4 reflects on the relationship between cultural workers, the suburbs, and the middle class. The main drawbacks to professional suburban cultural production are identified as the lack of cultural infrastructure, and the resulting isolation and stigma. These findings inform the next two chapters. chapter 5 critically analyses how cultural infrastructure is addressed in cultural plans and reveals repeated veneration of the cultural-district model of central cities in suburban cultural economic strategies. In advocating for a policy that moves beyond flagship suburban cultural boosterism, chapter 6 considers the value of investment in smaller-scale, multi-purpose, informal community service hubs and the flexible social networks that connect and animate them. Particular attention is directed to the role of cultural workers as suburban place-makers. Chapter 7 demonstrates how the place-making skills of suburban cultural workers are enhanced by computer-facilitated communication networks. Suburban cultural workers rely upon networks of association to productively extend their connections of creative labour beyond the immediate boundaries of the suburb. The suburban place-making exercise to which cultural workers are such valuable direct and indirect contributors remains an ongoing collective social project. Suburban places, like their creators, are in a constant state of becoming.

Chapter Two

The Geography of Somewhere:
The Suburbs

In the twenty-first century, the proportions of cities continue to change both demographically and spatially. The spillage of urban growth out from the edges has been accompanied by increasing popular and scholarly interest in suburbia and the process of suburbanism (Peck, 2011a). Suburbs no longer defy definition, nor are they studied with the intense disdain once afforded the "geography of nowhere" (Kunstler, 1993). Academic revisionists have shown that suburbia is "a place, as opposed to no place" and that it does not crush lives; instead it has the potential to enable them (Peterson cited in Kolb, 2008, 144).

This book is well served by a growing scholarly fascination with the historical development and contemporary reality of suburbs and suburban life. Important studies of suburbs have been undertaken within the American context by Kenneth Jackson (1985), Robert Fishman (1987), John Stilgoe (1988), John Teaford (1997), Dolores Hayden (2003), and Robert Lang (2003), and within the Canadian context by Richard Harris (1996; 2004), Larry Bourne (1996), Robert Lewis (2002), Carolyn Whitzman (2009), and Douglas Young et al. (2011). The epic stories told about the historical development of suburbia in this academic literature are overwhelmingly technical and practical in nature. The language employed by scholars documents technological, demographic, land-use, and policy changes that transformed the morphology of suburbia. Inspired by feminist critics (e.g., Hayden, 1984; Marsh, 1990; Weisman, 1994; Giles, 2004), who revealed the unpaid domestic labour, gendered lived realities, and masculinist planning ideology underpinning suburban built form, this book advocates for a reframing of the suburban narrative to consider the role that culture has played in the (re)development of suburbia. In support of such a cultural reframing, this chapter summarizes some of the conventional stories told about the foundation of the suburbs as a specific category of inhabited space (Vaughan et al., 2009). It then introduces the locations of the four inner and outer suburban case studies in the Toronto and Vancouver CMAs upon which this book is based.

A Brief Anglo-American History of the Suburb

The word *suburb* first appeared in the English language in the late fourteenth century (Girling & Helphand, 1994). In its original usage in medieval Europe, suburbs referred to the space outside, but adjacent to, the boundary of the city. Suburbs were physically and socially isolated places where taxes, moral laws, and political power did not apply and where people of marginal socio-economic status lived. Etymologically, *sub-* meant being near to, up against, or beneath. A *sub-urb* was thus interpreted as a sub-standard place "outside, beyond, or below the 'urb'" (Bourne, 1996, 164). From the outset, the word suburb described a relationship of proximity, dependency, and urban modification. A suburb was "a locale less than urban, at the fringe, and beneath the city in significance" (Girling & Helphand, 1994, 7). The etymological residue of subordination of the suburb relative to the city has lingered into the present day.

Located at the perimeter of cities, suburbs have historically marked the frontier of urban development. Each phase of suburban growth has been shaped by the input of theorists, policy makers, designers, developers, builders, realtors, journalists, and residents, as well as by the technological, resource, and political-economic contexts (Knox, 2008).[1] Within the North American context, the nineteenth century marks a useful starting point for initiating a suburban historical narrative.

In the late eighteenth and early nineteenth centuries, North American central cities were quite uncomfortable places – they were dirty, polluted, crowded, and disease-ridden. The outskirts of cities may have lacked the density of the core, but in the 1820s they housed noxious industries such as slaughterhouses and glue factories, and social reform establishments such as orphanages, hospitals, and prisons (Hayden, 2003). Neither urban nor rural, the edge of the city was a transition zone between city and country where the land-speculating motivations of developers collided with the pastoral fantasies of residents. Following in the footsteps of their British counterparts, a distrust of and dislike for the poverty, diversity, and evil influences of central-city urban life motivated many middle- and upper-class families in the United States and Canada to seek homes on the periphery of industrial cities in supposedly socio-economically homogeneous middle-class suburbs.

In her detailed case study of the historical development of Parkdale, once a politically independent suburb of industrial Toronto, Carolyn Whitzman (2009) turns the assumption of homogeneity on its head. She reveals the long history of diversity of people, household structures, housing, and land uses that characterized this now inner-city neigh-bourhood.[2] Working-class, lower-middle-class, and upper-middle-class families were attracted to the periphery of the city because land was cheap, houses could be built with sweat equity, and unspoiled nature was close at hand. As Whitzman (2009) details, developers frequently advertised their subdivisions as being on higher ground, close to nature, and upwind from smoke, and used place names with words such as *park, height, hill,* and *west* to convey these attributes to prospective buyers. To those individuals with secure employment, money, and leisure time, the middle-class Victorian suburb gave access to new homes with gar-dens on wider streets. This "suburban dream" of a detached house and yard, the stability of homeownership, and the supposed freedom and mobility afforded by the car was dependent upon an ideology of female domesticity, which advocated that women ought to bear and raise chil-dren, nurture a family, and create a tranquil domestic haven buffered from the stresses of work in the city.[3] The suburban dream also relied upon an ideology of division – territorial divisiveness and separation of work and leisure (Machor, 1987). By keeping unwanted land uses, activi-ties, and people out, the domestic ideal of the suburb as home to and protective of the virtuous values of middle-class women and children could be maintained. Thus the suburb came to represent progress, moral purity, physical health, and financial and residential stability; it was thought to be the solution to urban problems (Whitzman, 2009). This utopian "faith in the fresh landscape as a vehicle for creating a new urban milieu separate and different from the old" was the legacy of the nineteenth-century suburb (Machor, 1987, 212).

By the early twentieth century, Anglo-American culture seemed poised to embrace the idea of a suburban future. Private developers capital-ized on the demand for home ownership generated by growing urban populations with increased disposable incomes by constructing a range of suburbs from working-class to middle-class to exclusive.[4] Advances in transportation technology brought widespread reliance on the car as an increasingly affordable and dominant mode of travel, and an ensuing reconfiguration of urban space. The periphery became absorbed into the central city. What were suburbs for one generation of residents became indistinguishable from inner-city neighbourhoods for the next, as the urban-suburban-rural fringe boundaries were drawn ever outwards (Jindrich, 2010).

The decentralized suburban spatial fix was financially underwritten by significant multi-tiered government investment in infrastructure to reconstruct the central city and facilitate suburban expansion: road and highway routes; water and sewage service networks and treatment plants; and electricity and telecommunications grids. The road system, in particular, influenced the scale and pace of suburban development from the mid-twentieth century onwards. It "spawned large development parcels, single-use districts, and vehicular movement as a desirable activity, almost to the exclusion of transit and walking" (Sewell, 2009, 71). The expansion of the highway system in the 1950s and 1960s, combined with tax incentives and government subsidies for suburban real estate development, ensured that returning war veterans, new immigrants, and the rapidly growing population would be adequately housed and have access to reasonable home financing (Knox, 2008). A need for housing also stimulated the construction, automotive, and consumer durable industries, which, in turn, helped to underwrite mass suburbanization (Buck-Morss, 2000). The 1950s have since been characterized as "the largest suburban decade ever" because of the rapid population growth of the suburbs relative to the central city (Bourne, 1996, 171). The acceleration of suburbanization after the Second World War is attributed to the powerful legislative conjunction of the real estate, banking, and construction sectors (Hayden, 2003; Knox, 2005).

Throughout the long history of suburban development, the residential function of suburbs has predominated. That the original purpose of suburbs was to provide a living space for nuclear families was asserted through the 1950s and 1960s with the widespread construction of middle-class bedroom communities outside many Canadian cities (Halseth, 1996; Harris & Lewis, 2001; Harris & Forrester, 2003; Penfold, 2004; Sies, 2001). These dormitory communities of bungalow-style, split-level tract housing were initially thought to have little history and few social hierarchies. The American social critic Lewis Mumford (1961) famously and repeatedly denigrated suburbia as uninteresting residential wastelands of conformity, consumption, and mediocrity that were "bereft of cultural venues and activities of the mind" (Beauregard, 2006, 138).

Suburbia, however, is more than a homogenizing, domesticated residential "subtopia" (Knox, 2008). Through the 1960s, light industry, manufacturing, and higher-order retail followed residents out of downtowns to the suburbs, creating decentralized concentrations of jobs and services. Industrial parks and office parks were constructed, and trucking, with its reliance on the highway system, became the dominant mode of freight transportation. Private developer-builders constructed regional shopping centres and roadside retail strip malls, intensifying commercial land uses along major arterial roads and near highway exits.

Through the 1970s and 1980s, the suburban fringe continued to rapidly expand and transform as the suburban population surpassed that of the urban population in Canada (Bourne, 1996). In the outer suburbs, white-collar office space, high-income housing, large-scale mixed-use master-planned complexes, and high-rises were introduced to rival central cities (Filion and Gad, 2006). These suburbs developed downtowns of their own around multi-use retail, office, administration, and transit nodes (Filion & Gad, 2006). In the 1990s, suburbanization accelerated, turning metropolitan regions inside out.

In *A Field Guide to Sprawl*, an illustrated dictionary of suburbia, Dolores Hayden (2004) provides a general American audience with the vocabulary to describe and visualize the contemporary suburban landscape.[5] Glossy colour photographs accompany the text, using a bird's-eye perspective to create monumental landscapes. In these images the distended architectural scale of generic buildings designed with a logic that values short-term efficiencies is contrasted with a vast, horizontally built, and repetitious landscape that is punctuated by "ambiguous in-between spaces" and "vast zones of indifference" (Ingersol, 2006, 76). On the ground, such proportions can unsettle and isolate people, hiding the lived realities of places from view (Wright, 2008).

The Name Game

At the close of the twentieth century and into the twenty-first century, suburbs have flourished as objects of scholarly and popular study. American academics and journalists alike have stumbled over themselves to develop a new vocabulary of suburbanization. Neologisms abound in suburban studies: "technoburb" (Fishman, 1987); "edge city" (Garreau, 1991); "hundred-mile city" (Sudjic, 1992); "exopolis" (Soja, 1996; 2000; 2008); "edgeless cities" (Lang, 2003); "boomburb" (Lang & Lefurgy,

2007); and "metroburbia" (Knox, 2008). All of these forms of suburban settlement are associated with urban decentralization and sprawl and have been classified by Nicholas Phelps et al. (2006) as "post-suburban." This scholarly classification game illustrates an apparently unending quest to claim intellectual territory.

In his book *Bourgeois Utopias*, Robert Fishman (1987) uses the term "technoburb" to describe decentralized suburban cities that are made viable as "socio-economic units" by communication technologies and networks of highways that blur the boundaries between work and home. The technoburb is said to provide all of the amenities necessary to sustain work and life – shopping malls, industrial parks, office complexes, hospitals, schools, and a range of housing types. Focusing on the outward migration of office parks and shopping malls and the rapid growth of suburban cities on the periphery of city-regions, American journalist Joel Garreau (1991) introduced the term "edge cities" into common usage. In his landmark book *Edge City: Life on the New Frontier* Garreau documents how American suburbs have silently materialized as employment and retail hubs for their regions, assuming a role once reserved for central cities.[6]

Inspired by Garreau's work, Robert Lang (2003) declares that not all suburbs have the density and cohesiveness of edge cities. For the free-form office development that sprawls along highways and highway interchanges, filling out the interstices between central cities and more concentrated suburban business districts, Lang employs the phrase "edgeless city." In collaboration with Jennifer Lefurgy, Lang (2007, 73) asserts that the "boomburb" is "the new face of the American suburb." Boomburbs are understood to be suburban jurisdictions with populations over 100,000 that have sustained double-digit rates of population growth in the three decades between 1970 and 2000.[7] Using this definition, both Mississauga and Surrey can be considered Canadian examples of boomburbs.[8] Lang and Lefurgy (2007) argue that boomburbs are as large, influential, and ethnically heterogeneous as more established cities with longer histories, but that they are generally unknown and unremarkable. As exemplars of accidental urbanism, boomburbs are said to possess many of the same components as cities but lack dense business cores and a high-quality public realm. Instead, boomburbs are characterized by a fast-growing, low-density, car-dependent, and minimally ordered spatial structure. The term boomburb references the generational descriptor "baby boomer," the demographic bulge after the Second World War. By sheer force of numbers, baby boomers are considered to have remodelled society as they moved through it, leaving one to question whether boomburbs, like their generational namesake, are really as numerous, unique, or as change-instigating.

Baby boomers with distinguished academic careers have also been heavily involved in the ongoing suburban labelling exercise. In the mid-1990s, with Los Angeles as his urban research laboratory, Edward Soja (1996, 238) coined the term "exopolis" to describe the "oxymoronic ambiguity" of the "amorphous implosions of archaic suburbia." Exopolis is meant to convey the exogeneous forces shaping "outer" cities and the rise of "ex-cities" that lack the "traditional traits of cityness" (Soja, 2008, 148). With boundaries difficult to delineate, these are places with landscapes characterized by spectacle, simulacra, and consumption. This "leading edge of new development" in American suburban and exurban areas, where "residential settings ... are thoroughly interspersed with office employment and high-end retailing," Paul Knox (2008, 5 and 38) refers to as "metroburbia."

The new language of suburbanization that I have touched on in this chapter aims to capture some of the ways in which the American city-region in the twenty-first century has been dramatically "stretched and reshaped to accommodate increasingly complex and extensive patterns of interdependency" (Knox, 2008, 37) that are not adequately conveyed in the simplistic city-suburb binary. It is a language that focuses on the restructuring of American (sub)urban physical form and economic function. It is a language driven by economic and planning agendas that attends to shifts in labour markets, morphology, and land uses, but largely neglects lived social experiences and the transformative force of culture. It is a language that fixates on new shapes, patterns, and forms at the furthest suburban reaches of the decentralized region and, in so doing, overlooks the complexities of older, inner suburbs.

The Shifting Fortunes and Possibilities of the Inner Suburbs

Built from the mid-1940s until the 1960s, "older," "mature," "postwar," "first-tier," "inner-ring," or "inner" suburbs, as they are variously referred to in the North American professional urban planning and scholarly urban studies literatures, are located just beyond the core of the central or mother city, yet are contained within its urban orbit (Clapson, 2003). Joel Kotkin (2000, 35) labels these older suburbs "the midopolis" and describes them as "less a frontier of development than a shifting middle ground between the urban core and new growth nodes along the metropolitan edge." Often built as extensions to, or amalgamations of, central cities, they are an ambiguous in-between landscape where inner and outer city mix (Soja, 2008). They are a "middle way, lying somewhere between the extreme, polarized conditions of urbanity and the rural" (Dault, 2001, 10). There are numerous interdependencies as the central city is commutable and often remains a necessary source of goods, services, employment, and entertainment for inner-suburban residents (Greenstein & Wiewel, 2000). However, there are also many inner-suburban residents who are employed locally or regionally. Mattingly (1997) suggests that older suburbs were stabilized by families who remained in the same locale for generations. A common criticism of inner suburbs is that they lack a physical, public centre (Martinson, 2000). The architect Avi Friedman (2002, 52) proposes that the provision and maintenance of community centres would be one way "to feel unified as a community for what may otherwise be an unfocused place." He recommends that in residential suburbs a centre point of activity could also be a provided by a high school, a prominent place of worship, or a recreation centre. Community focus also sometimes comes from privatized public spaces (e.g., the food court) in neighbourhood shopping plazas and regional shopping malls – although the "dead mall" syndrome and the creation of greyfield space would suggest that small local shopping malls may no longer be a financially sustainable built form in the inner suburbs (Parlette & Cowen, 2011).[9]

Where the central city is commonly a destination, the inner suburbs are rarely viewed by urbanites in this light. Instead, inner suburbs are usually considered places that "auto-flaneurs" drive *through* rather than *to* (Husler, 1997). Compared with their central-city counterparts, inner-suburban neighbourhoods tend to have fewer non-profit organizations, mixed uses, and recreational, entertainment, retail, and public transit options (Lucy & Phillips, 2000). Activities are often functionally segregated and require private modes of transport to access. The suburbs were designed for exclusivity and separation, and it is frequently asserted that they do not "add up to a coherent system of living, working, playing, school, and shopping" (Marshall, 2000, 105).

Where suburbs were once considered by the nineteenth-century bourgeoisie to be an escape from the congestion, pollution, crime, and social ills of the inner city, they now face many of the same "problems" as central cities (Mattingly, 1997). When inner suburbs enter the media spotlight, it is usually to highlight the extent of (sub)urban decay and lawlessness within their boundaries rather than to highlight their civic potential and achievements.[10] Certainly inner suburbs have their fair share of challenges. They are underserviced and "they have become run-down and unfashionable, losing wealth and population" (Thomas, 1998, 113). In William Hudnut's (2003) book *Halfway to Everywhere: A Portrait of America's First-Tier Suburbs*, inner suburbs are described as possessing "aging infrastructure, deteriorating schools and commercial corridors, and inadequate housing" and as having a disproportionate number of immigrants, working poor, single-person households, and aging original homeowners as residents. The Canadian context is not dissimilar, and debates about it have gone mainstream.

In the run-up to Toronto's 2010 mayoral election, a newspaper article highlighting talking points for candidates suggested that Toronto with its reverse white-flight to the central city is rapidly becoming an "inside-out version of Detroit" because "it has not yet figured out a way to bring wealth outside its core to the suburbs which are increasingly blighted by poverty, crime, and disrepair" (Friesen et al. 2010, M1). A report issued by the United Way (2004), *Poverty by Postal Code*, determined that the number of high-poverty neighbourhoods increased from 30 in 1981 to 120 in 2001, with 92 of these concentrated in the inner suburbs. In 2005, the City of Toronto in conjunction with the United Way of Greater Toronto identified 13 "priority neighbourhoods" (areas of the city designated in need of financial investment because of high poverty rates and few social or community services); all of these are located in the inner suburbs. Building on these findings, David Hulchanski (2007) uses data on change in average individual income over three decades to illustrate the emergence of a city spatially divided along income and ethno-cultural lines. The postwar suburbs, Hulchanski (2007, 9) reiterates, have the greatest concentration of people with low incomes, especially newly arrived immigrant families housed in high-rise apartments, living in "an urban landscape with a 30-year history of abandonment by people who have a choice." The authors of all of these studies acknowledge that these findings should come as no surprise given that a comprehensive study conducted by the Social Planning Council of Metropolitan Toronto shed light on the problem back in 1979. A follow-up policy report to this study, entitled *Planning Agenda for the Eighties – Part II: Metro's Suburbs in Transition*, detailed the necessity for civic leaders to assume greater responsibility for addressing the emergent socio-economic needs of growing numbers of single parents, newcomers, unemployed, and youth in the suburbs.

Social science research confidently supports the claim that in contrast to the white, middle-class, and family-oriented stereotype that remains heavily marketed to the public by developers, North American inner suburbs are socially diverse places, with population bases reflecting a variety of income and educational levels as well as ethnic backgrounds (Vachon et al., 2004; Ray et al., 1997; Hall & Lee, 2010). Additional characteristics of Canadian inner suburbs include housing stock that is in disrepair and continues to deteriorate, minimal public and private commitment to infrastructural investment, high poverty levels, family income decreases (brought about in part by a loss of manufacturing employment), crime rate increases, and tax base erosion.

Other scholars put a more positive spin on the situation, suggesting that deteriorating inner suburbs should not be categorically framed as "grey areas" of degradation, devaluation, and disinvestment (Kendig, 1979, 3). David Listokin & Patrick Beaton (1983, xviii) emphasize that the infrastructure of older suburbs is competitive; housing, for example, is "available at prices to which moderate-incomed people can aspire, and thus can be a crucial stepping stone toward full middle-class status for those who cannot afford the full costs of fashionable suburbia." But in many Canadian inner suburbs the residential landscape is changing. The working-class bungalow, whose construction was sponsored by the Canadian Housing and Mortgage Corporation in the 1950s and 1960s, has been replaced by stucco houses with disproportionately large footprints, themed townhouse developments, and condominium towers.

Inner suburbs, because of their proximity to the central city, renovation and reinvestment potential, and diverse demographic composition, are in a unique position to exert a positive influence on urban development (Filion et al., 2011). David Listokin and Patrick Beaton (1983, xviii) characterize older suburbs as both "the forgotten frontier of America" and "the new frontier of our society." City managers are rarely so visionary or inspired. Instead, urban decision makers frequently disregard inner suburbs "as insignificant pieces in the metropolitan mosaic" (Hudnut, 2003, 91). They do so precisely because of the extent of decline that began in the 1980s, the concentration of socially and economically marginal and powerless populations, and the potentially high municipal outlays (Lucy & Phillips, 2000).[11]

Older suburbs, then, need to be forcibly pulled out of the "policy blind spot between the attention long directed to central cities and new attention focused on fast-growing exurban areas" (Hudnut, 2003, ix). To add value to inner suburbs is to appreciate that they are works in progress which are not yet fully understood (Noble, 2009). Hudnut (2003, 109) advocates that policy makers "work to discover ways to unleash human potential and forge ties of social capital that bond people together in a variety of networks and bridge the differences that can easily divide them." This is an admirable and significant goal that I, too, strongly support. However, I would extend Hudnut's list of resources (e.g., local government, regional non-profit organizations, faith-based institutions, and universities) to include cultural workers. I maintain that central to the restorative, egalitarian, and sustainable development of inner suburbs is the utilization of the grassroots knowledge and expertise of local cultural workers. Suburban cultural workers are often deeply embedded in communities where they work closely with local residents, celebrating local histories, and building local networks. In much the same way that Dolores Hayden (2003, 245) has described suburbia, cultural workers can also be understood as "the hinge, the connection between past and future, between old inequalities and new possibilities."[12] Subsequent chapters explore in greater detail the reflexive and recursive relationship between cultural workers and inner and outer suburbs. The remainder of this chapter, however, introduces the locations of the primary suburban case studies upon which this book is based.

Toronto: From Etobicoke to Mississauga

Data from the 2006 census show that 13.9 million people, nearly 50 per cent of the Canadian population, live in the country's three largest city-regions: Toronto, Montreal, and Vancouver (Statistics Canada, 2007). These CMAs possess some unique demographic attributes: their population growth is due primarily to the immigration and in-migration of young adults, and their surrounding suburban municipalities have larger populations and population growth rates than other CMAs in Canada. Of the 25 municipalities with the largest populations in Canada, nine of these are located on the suburban fringes of Toronto, Montreal, and Vancouver, with particular mention made of Mississauga and Surrey (Statistics Canada, 2007).

Once a relatively dense and compact city in the mid-twentieth century, Toronto has since absorbed farmland to become the most economically powerful, populated, and heavily urbanized city-region in Canada. Bounded by Lake Ontario to the south, the Kawartha Lakes to the east, Lake Simcoe to the north, and the Niagara Escarpment to the west, the Toronto CMA sprawls in a linear pattern along the western shores of Lake Ontario. Throughout its history, economic, political, and urban restructuring has extended the boundaries of the city and the city-region. In 1996, the provincially mandated City of Toronto Act forced the cities of Etobicoke, North York, Scarborough, York, and Toronto and the Borough of East York to amalgamate into a megacity. In the contemporary period, the City of Toronto and the surrounding regional municipalities of Halton, Peel, York, and Durham together form the Greater Toronto Area (GTA) (Figure 2.1). The area includes 25 municipalities and four regions across 7,000 square kilometres. The GTA contains 5.6 million people and receives over 150,000 new residents each year. Much of that population growth, in conjunction with increases in business and residential construction, has occurred on the suburban fringes of the city-region.

Incorporated as a township in 1850, Etobicoke merged with the three lakeside municipalities of Mimico, New Toronto, and Long Branch in 1967 to form the Borough of Etobicoke before becoming the City of Etobicoke in 1984. With amalgamation in 1998, it lost its status as a separate municipality and publicly assumed the identity of an inner suburb of Toronto. Etobicoke now contains about 13 per cent of the population of the City of Toronto, although the number of Etobicoke residents, half of whom are immigrants, has declined in recent years from 338,117 in 2001 to 334,491 in 2006. Covering 124 square kilometres (about 20 per cent of the City of Toronto's land area), Etobicoke extends from Lake Ontario north to Vaughan and is bordered on the west by Mississauga and on the east by the Humber River, but it has no clearly defined centre. It is fractured into three districts – north Etobicoke, central Etobicoke, and south Etobicoke – that span the socio-economic spectrum from extreme wealth to overt poverty, with very little in the middle.

In the 1980s, north Etobicoke (Rexdale) was a middle-class manufacturing centre. Through the recession of the 1990s, north Etobicoke lost many of its major employers and became a place that the news media strongly associate with the negative forces of urban decay, gun violence, drugs, and gang activity (LeBlanc, 2012; Winsa & Rankin, 2012). It is a neighbourhood of residential high-rises (the largest concentration in Toronto), subsidized housing, and a large immigrant population.[13] North Etobicoke was labelled a "priority neighbourhood" in former Mayor David Miller's community plan and has been targeted by municipal leaders for revitalization through projects that combine culture and recreation. While Rexdale may be portrayed as a "forgotten fringe of failed factories, threadbare strip plazas, towers-in-the-park and a smattering of tidy 1960s bungalows," it is also a place with intriguing examples of adaptive reuse and low-risk entrepreneurial and homeownership opportunities (LeBlanc, 2012, G4).[14]

To the south, central Etobicoke is a predominantly white, upper-middle-class neighbourhood of old homes and old money centred on

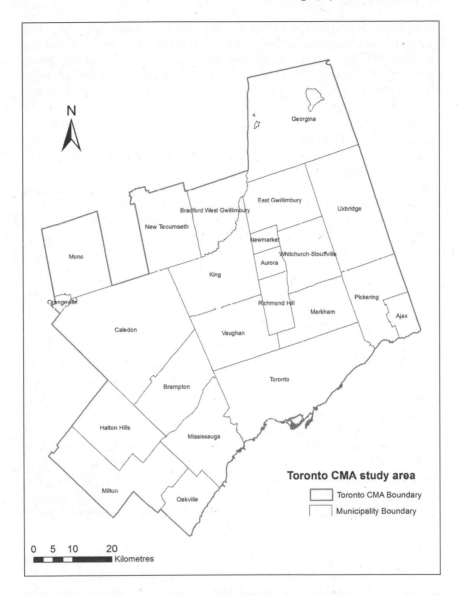

Figure 2.1   Municipalities and study area within the Toronto CMA (map by W. Sturm).

Bloor Street West and the Kingsway. Further south still, south Etobicoke (The Shore) is a geographically isolated, east-west sliver of land between the lake and the factories and transportation corridors to the north made up of three neighbourhoods, Mimico, New Toronto, and Long Branch, with long histories as towns in their own right. In the 1890s, Mimico was where some of Toronto's wealthy industrialists built their summer homes; New Toronto was a heavy-industry waterfront community built around the railway line, with housing for factory workers; and Long Branch was a summer resort town of cottages, hotels, amusement rides, and a boardwalk. Each retained town status until 1967. Today, south Etobicoke is ethnically diverse, with high numbers of Korean, Spanish, Polish, Lebanese, Caribbean, and East African residents. While its population remains predominantly working-class and contains the largest concentrations of cultural workers in Etobicoke, there are also pockets of middle-class wealth in older homes along the shoreline and in new condominium and townhouse infill.

To the west of Etobicoke is Mississauga. In 1974, the towns of Mississauga, Port Credit, and Streetsville, as well as parts of the townships of Toronto Gore and Trafalgar, were amalgamated into the City of Mississauga. Despite amalgamation, Mississauga remains divided into 11 wards and some residents maintain that these wards continue to act like small villages rather than effectively collaborating together as a coherent and cohesive city. For over three decades, Mayor Hazel McCallion has led Mississauga's local government.[15]

Since incorporation, the City of Mississauga has grown rapidly to become the sixth-largest city in Canada and the third-largest city in Ontario. With a population in 2011 of over 710,000, much of this growth is attributed to immigration (Preston, 2002). According to the 2006 census, just over half of the population are foreign-born, with 50 per cent from Asia and the Middle East, 30 per cent from Europe, and 7 per cent from the Caribbean. There are vibrant Polish, Chinese, Punjabi, Portugese, and Italian communities in Mississauga, and over 90 different languages are spoken.

Mississauga's population is aging (45 per cent of the population is in the 35–50 age group) and over 14 per cent of the population lives in poverty (Lorinc, 2008). In addition to meeting the needs of a diverse and aging population, city managers recognize the need to retain and attract young people to the city. Mississauga's workforce is well-educated, with 60 per cent holding post-secondary degrees; of those, 5 per cent have post-secondary degrees in fine and applied arts. In 2006, 1 per cent of Mississauga's labour force were employed in professional occupations in arts and culture. The residential districts with the largest numbers of people so employed are located along the shores of Lake Ontario in southern Mississauga, in Lakeview, Port Credit, Clarkson-Lorne Park, Sheridan, and Erindale.

Provoked into being by the construction of the Square One shopping mall and its framing highway corridors, Mississauga has long been characterized as a development-happy place (Hume, 2012). There are no height restrictions in Mississauga's downtown. With its towers of brick and glass Mississauga has become one of the densest suburban communities in North America (Rochon, 2007). A concentration of head offices and pharmaceutical and electronic companies has made Mississauga a place to which many Torontonians now reverse-commute for work. Mississauga has approximately 55,000 registered businesses, employing more than 425,000 people, and is home to 61 Fortune 500 Canadian or major divisional head offices and 50 Fortune Global 500 Canadian headquarters (http://www.mississauga.ca/portal/discover/aboutmississauga).

Mississauga's moniker, "leading today for tomorrow," and its street banner branding campaign, "moving forward," convey a sense of the rapid pace of (sub)urban change and the desire to be at the forefront of it. Yet the city is quickly becoming built out. The revenue generated from development charges is decreasing, existing infrastructure is aging, and new transit and cultural infrastructure is needed. City councillors have spent the last half-decade looking to international consultants for urban planning direction to help transform Mississauga "into a city in its own right with its own authentic identity" (City of Mississauga Culture Division, 2009, 28). In an article titled "Could Mississauga Ever, Ever Be Sexy?" architecture critic Lisa Rochon (2007) describes "electrifying, high-volume civic boosterism" at work in Mississauga. "A Conversation about Building a City for the 21st Century," a weekly speaker series with "urban visionaries" held during November 2007 at the Mississauga Living Arts Centre, sought to engage thousands of residents in the city-building process. Despite the many ideas proposed, the process of place-making in this young city follows the formula employed by other municipalities, namely, downtown intensification through condominium development, starchitectural branding, waterfront revitalization, and investment in a privatized public realm. Take, for example, the recently completed 56-storey Absolute World luxury condominium towers designed by Beijing architect Yansong Ma and nicknamed the Marilyn Monroe and Monroe Mate because of their curvaceous forms (Figure 2.2). These towers overlook the city centre and are intended to be

Figure 2.2   Mississauga's Absolute World condominium towers, Marilyn Monroe and Monroe Mate (photograph by L. Wheelans).

a "landmark" that contributes to a distinctive Mississauga skyline (www
.condosky.com/en/luxurycondos/theAbsoluteWorld.html).   Journalist
Frances Bula (2011) asserts that Mississauga is "the most advanced of
the suburban downtown efforts anywhere in Canada, the town centre
is still dominated by a large regional mall, Square One, several grand
new civic buildings constructed a quarter century ago (the kind of building boom that Surrey is undertaking now), and only the beginnings of
some destination architecture." In the contemporary period, as through
much of its history, Mississauga residents and decision makers continue
to oppose the idea that Mississauga is little more than a suburb or a
sleepy bedroom community of Toronto; they continue the struggle to
craft a unique identity for their city, distinct and separate from Toronto.

Vancouver: From North Vancouver to Surrey

The Greater Vancouver Regional District (GVRD) in the southeast corner of British Columbia's Lower Mainland emerged in 1967 as a product of provincial regional district legislation. In 2007, the GVRD was
renamed Metro Vancouver. It is a federation of municipalities and elec-
toral areas that covers nearly 3,000 square kilometres, just 4 per cent
of the province's land area, where the Fraser River meets the Strait of
Georgia and the Pacific Ocean (Artibise & Meligrana, 2003).[16] The area
is surrounded by mountains on the north and east, the sea on the west,
and the American border on the south and these natural and political
boundaries have limited the size and direction of urban sprawl. (Sub)-
urbanization has extended linearly in an easterly direction inland up
the Fraser River Valley. Metro Vancouver now contains nearly half of
British Columbia's population within 21 municipalities (twelve cities,
six districts, and three villages) (Figure 2.3). The population of the
Vancouver metropolitan region has increased from 890,000 in 1961 to
2,116,581 in 2006, with suburban municipalities surrounding Vancouver
absorbing a significant share of the population growth. Between 2001
and 2006 the rates of population change have been most rapid for
municipalities in the eastern and southern parts of Metro Vancouver:
Maple Ridge +9.2 per cent; Langley +6.1 per cent; New Westminister
+7.1 per cent; Port Moody +15.5 per cent; and Surrey +13.6 per cent
(Statistics Canada, 2007). Within the regional institutional framework of
Metro Vancouver, individual municipalities retain their independence
from one another and have developed inter-municipal governance structures on an ad hoc and incremental basis (Artibise & Meligrana, 2005).[17]

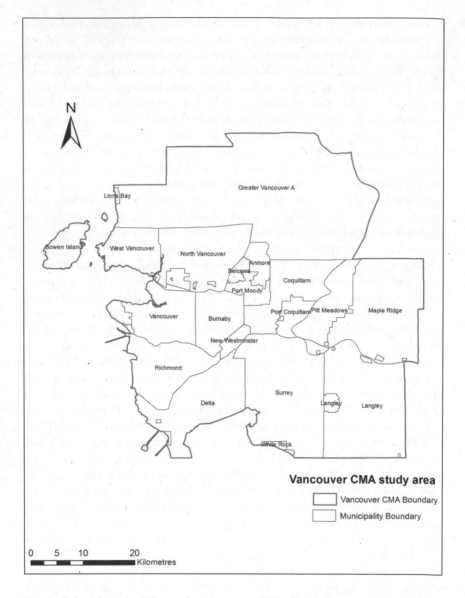

Figure 2.3   Municipalities and study area within the Vancouver CMA (map by W. Sturm).

Across Burrard Inlet, at the foot of the Coast Mountains and surrounded on the east and west by the suburban-like District of North Vancouver, the City of North Vancouver developed as an important centre for lumber, milling, shipbuilding, and transportation. The relatively compact city of only 12 square kilometres supported a population of 45,165 in 2006 (an increase of only 862 residents from 2001) and provided workspace for 23,000 employees. Only 30 per cent of North Vancouver residents actually work in the municipality where they live (Metro Vancouver, 2008a). The top three areas of employment in the City of North Vancouver are sales and service occupations (27 per cent), business, finance, and administration occupations (19 per cent), and management (12 per cent) (City of North Vancouver, 2009a). Employment in arts, culture, recreation, and sport registered at nearly 6 per cent of the workforce in both the city and the district (Metro Vancouver, 2008b).

The city is composed of nine neighbourhoods, with the majority of residents concentrated in Lower Lonsdale and Central Lonsdale in a five-block-wide corridor that is bisected by Lonsdale Avenue and runs south to the waterfront. These two most southern and central residential neighbourhoods are particularly attractive to cultural workers, seniors, single-parent families, and immigrants from Britain, Iran, the Philippines, China, and South Korea. Iran remains the most common place of birth for North Vancouverites, although between 2001 and 2006 the proportion of Iranian immigrants decreased from 34 per cent to 26 per cent of residents, and the number of immigrants from the Philippines and South Korea increased. The 2006 census shows that English (61 per cent) remains the dominant language in North Vancouver homes, followed by Farsi (6 per cent), Chinese (3.4 per cent), Korean (2.3 per cent), and Tagalog (2.1 per cent).

Near the turn of the twentieth century, economic prosperity in the Lower Lonsdale area of North Vancouver led to its separation from the District of North Vancouver. Since incorporation of the City of North Vancouver in 1907, the waterfront has remained an important focal point for economic activity and redevelopment. Lonsdale Quay Market is now a significant retail and transportation hub for buses and the Seabus ferry, which provides a 15-minute boat commute to downtown Vancouver. In an ongoing effort to transform the waterfront into an urban destination for residents and tourists, the municipality has collaborated with a private developer to restore the former historic shipyard site.[18] With its densifying residential neighbourhoods and large-scale waterfront redevelopment initiative, the City of North Vancouver wants

to be the "urban core of the beautiful North Shore" (http://www.cnv
.org/). While municipal leaders may prefer the label urban to inner
suburban, there is no doubt that the City of North Vancouver exists in
a dynamic, yet deferential, relationship with the City of Vancouver. It
retains strong socio-economic, political, cultural, and transportation ties
across Burrard Inlet, as well as with the two other municipalities on the
North Shore. Certainly throughout the history of the North Shore many
facilities and services have been shared between the municipalities. The
rail, bridge, road, and ferry links across to Vancouver have been essential to North Vancouver's economic survival. Most Vancouverites seldom
think of the North Shore as comprising individual municipalities.

The two North Vancouvers, the City and the District, with their separate mayors and councils, have long been envisioned by their residents as a collection of villages rather than as discrete, singular places.
Although technically they are two separate municipalities, most Lower
Mainlanders refer to the two "North Vans" as North Vancouver. In many
respects, the City of North Vancouver, with its higher commercial and
residential densities, acts as an urban core for the North Vancouver
District Municipality, which is more rural and suburban. Where the City
of North Vancouver's land supply is nearly built out and much new development is accommodated through infill projects, the District of North
Vancouver extends east-west across 160 square kilometres of steep roads
and forested slopes (Miller & Cavens, 2008). The majority of the District's
82,562 residents live in single-family homes in six neighbourhoods, each
with its own small commercial centre of locally owned businesses and
community services. From west to east, these neighbourhoods include
Edgemont Village, Upper Lonsdale, Lynn Valley, Main Street, Parkgate,
and Deep Cove.

Most development (e.g., the Cleveland Dam, the Lions Gate Hospital,
Capilano College, municipal halls, shopping malls, recreation centres,
libraries, and new subdivisions) occurred in the decades after 1950 as
the City and District began to recover economically and car ownership
increased. Much of this development occurred in the District, where
land was available and made accessible through investment in transportation infrastructure: namely, the opening of the Second Narrows
Bridge (the main link to Vancouver that can handle the heavy truck
traffic unsupported by the Lions Gate Bridge) and the construction of
the Upper Levels (Trans-Canada) Highway facilitating east-west linkages
along the North Shore. In the 1960s, the lumber and shipbuilding activities that once dominated the waterfront were replaced with new docks,

grain elevators, and terminals for shipping potash, fertilizer, and coal through an elaborate ship-rail-truck transportation system (http://www .northvanmuseum.ca/local4.htm). In the last three decades, the marine-focused economy has been supplemented by growth in high-technology and cultural industries (e.g., film and television studios) as well as by investment in recreational tourism.

Farther south and east of North Vancouver, the City of Surrey is located in the centre of Metro Vancouver in British Columbia's Lower Mainland. Founded on logging and agriculture, and formerly a district municipality, Surrey was incorporated in 1879 and obtained city status in 1993. The construction of a highway network in the Lower Mainland in the 1940s, combined with the lower cost of housing, helped to transform Surrey from a rural community into a bedroom community of Vancouver. The 1965 draft Community Plan developed by Louke Kleyn encouraged the development of five "towns" (Guilford, Whalley, Cloverdale, Newton, and Sunnyside, with the later addition of South Surrey as a sixth) focused around residential neighbourhoods and supplemented by community and recreational centres, shopping malls, green space, and farmland. These six town centres are incorporated into one city. In practice, however, they often function as separate cities, each with its own distinctive place identity based on immigration patterns and historical development.[19] The City Centre of Surrey is located in the community of Whalley near the Surrey Central SkyTrain station and hosts the Central City Shopping Centre and the Surrey campus of Simon Fraser University. Surrey Central has been designated as the new downtown core of Surrey, where city hall will be moved and where tax breaks will be offered to attract new office, condominium, and retail development (Bula, 2011).

Across Surrey, commuter houses, strip malls, office parks, motels, storage yards, and warehouses have replaced tracts of farmland. This "suburbification" of a once-agricultural landscape can be seen along many of the major traffic arteries, where new middle-class subdivisions surround working farms (Brayshaw, 1998). In Surrey, 35 per cent of the land base is classified as agricultural. With its greenhouses and vegetable cooperatives growing most of British Columbia's onions, carrots, celery, and lettuce, Surrey is considered the vegetable capital of the province. Today, however, rather than actively working the land, most Surrey residents drive past agricultural fields, appreciating them perhaps only as a visual trace of a slowly disappearing lifestyle (Brayshaw, 1998).

Surrey has the largest land area of any city in British Columbia at 317 square kilometres. It can take an hour to travel from one end of

Surrey to the other on a street system based on a rural, one-acre grid pattern (Bula, 2011). With a population of 395,000 in 2006 (up 13.6 per cent from 2001) and approximately 12,000 new residents a year, the city is expected to surpass the City of Vancouver (population 578,000) in population in a decade. In terms of population, Surrey is the tenth-largest city in Canada, yet nearly 50 per cent of the population in each of the town centres is employed outside Surrey, treating the municipality as a bedroom community (City of Surrey, 2009b). The three largest occupational clusters in Surrey are sales and service (25 per cent), trade, transport, and equipment operators (19 per cent), and business, finance, and administration (17 per cent) (City of Surrey, 2009b). Employment in art, culture, recreation, and sport represents 2 per cent of the local labour force.

The motto on Surrey's coat of arms attests "progress through diversity." According to the City of Surrey's Planning and Development Department (2009b), in 2006, 90 per cent of residents had Canadian citizenship and nearly 40 per cent of the population were immigrants, with the most common places of birth being India (35 per cent) and the Philippines (8.4 per cent). After English, Punjabi was the most common first language learned and spoken at home by Surrey residents. Many immigrants to Surrey have been refugees, forcing the city to prioritize settlement issues (particularly poverty, language, and educational barriers).

Despite the city's motto and the demographic statistical profiles available on the city's official website, it is the economic development potential of Surrey rather than the ethnic diversity that is highlighted there. The welcome message from Mayor Dianne Watts, on behalf of city council, emphasizes Surrey's strategic location on the American border, its diverse recreational landscapes of parks, beaches, and golf courses, its arts and culture, and its fresh agricultural produce. Over the last decade, city officials have worked hard to counter public perceptions of Surrey as a place that is rife with crime, poverty, racialized tensions, and ethnically segregated residential patterns.

Conclusion: Telling that Different Story

Each of the pairs of inner and outer suburbs on the outskirts of Toronto and Vancouver introduced in this chapter demonstrate that suburban places are "somewhere" to many Canadians. They each have unique histories, landscapes, and demographic profiles. And they each allow different lived experiences to unfold within them. Despite the particularities, there remain, as Dolores Hayden (2003, 17) has noted, many "spatial conventions and social expectations of the nineteenth and early twentieth centuries … tangled in memory and manners." These influence the identities, political and moral values, and social distinctions communicated through contemporary suburban landscapes. While the present is inevitably rooted in the past, it is time to tell a different story that (re)imagines the creative and cultural possibilities of suburbia.

A useful starting point for rethinking and rewriting the suburbs is John Lorinc's (2006) term "cultureshed." A play on the concept of watershed, the term cultureshed refers to the wide reach of culture needed to nourish a city-region's creative resources. The concept of cultureshed challenges understanding culture as an urban phenomenon. It extends the reach of culture into the suburbs and beyond into the mid-sized cities and smaller towns that punctuate the cultural hinterland (Bain & McLean, 2012; Bain & McLean, 2013). These are places that are often sacrificially devalued in the constructivist lexicon of the creative, cosmopolitan city. It is to the myths of bohemian marginality and the lived realities of suburban cultural production that discussion now turns.

Chapter Three

(Un)usual Suburbanites? Explorations and Interventions at the Cultural Frontier

This chapter employs the metaphors of exploration, frontier, and pioneer to interpret the postwar suburban landscape and to consider why artists are more widely represented as urban pioneers than suburban pioneers. It focuses on the reconfiguration of the suburban home as a site of cultural reproduction and artistic engagement. The work of American artist Gordon Matta-Clark is used as conceptual inspiration for interpreting the artistic interventions of the Calgary-based art collective the Arbour Lake Sghool (the spelling originated with the lack of "Cs" in a peel-and-stick kit used for an early promotional sign). This contemporary Canadian suburban art collective has, since 2003, played with residential disturbance and intervention as tools to question suburban domestic landscapes. With the work of the Arbour Lake Sghool as the discursive focal point, this chapter considers two questions. First, can artists be committed suburbanites? Second, can neo-bohemia be found outside the central city in a suburban cul-de-sac?

The Pending Suburban Ascendancy:
From School Bus to Disconnected Arts Audience

In the still-unfolding history of urban development, some scholars claim that the "suburbs belong at centre stage" (Kruse & Sugrue, 2006, 1). Despite the documented rise and dominance of suburbia in North America and the claim that the suburbs are "the most significant landscape to emerge in the twentieth century" (Beuka, 2004, 2), for central-city residents practical engagement with the suburbs is often limited. Urban residents may encounter the suburbs on a drive to Costco to stock up on super-sized boxes of cereal and styrofoam trays of rib-eye steaks or to the big-box clothing outlets. The slightly more adventurous

Figure 3.1   A cultural journey on a school bus from downtown Toronto through Mississauga to an art opening at the Heritage Complex in downtown Brampton (photograph by A. Bain).

day-tripping urbanites may take public transit, ride a bike, or walk beyond the retail strip. In Toronto, they may join a Jane Jacobs walking tour of Scarborough or an awareness-raising bicycle ride to York Region. They may take a free return school bus trip – from the downtown cultural lynchpins of the Gladstone Hotel or the Ontario College of Art and Design – to exhibition openings at the University of Toronto Mississauga Art Gallery or the Heritage Complex in Brampton (Figure 3.1).

Among them, the more *au fait* academics, urban planners, architects, journalists, artists, and municipal politicians (seeking re-election) may attend the opening of an exhibition at the downtown Toronto Design Exchange entitled *Fringe Benefits: Cosmopolitan Dynamics of a Multicultural City* (2008). Wine glass in hand, squeezing between bodies, they may seek to explore Toronto's inner suburbs "as something other, something distanced, a critical target understood through ... photographs" (Mattingly, 1997, 40–1). Like rows of identical houses in subdivisions, white foam-core placards hang from the gallery ceiling in virtual boxed neighbourhoods, each containing lush colour photographs, neat maps, and elegantly scripted descriptions of architectural, planning, and activist projects positioned as creative, innovative, and visionary: a Muslim subdivision in Vaughan; an indoor soccer field in Brampton; a cultural multiplex community centre in Markham; a Punjabi strip mall development in Mississauga; a Japanese-Canadian cultural centre in Don Mills; and a high-rise complex in Thorncliffe Park. While this exhibition is a necessarily selective, static, and scripted exploration of the ethnoscapes of Toronto's inner suburbs, it demonstrates emergent public recognition, among a self-selecting, educated, and politically motivated audience, of the significance and potential of the suburbs for future sustainable urban development initiatives (Chodikoff, 2008).

Exhibitions and interventions can be important tools to initiate critical engagement with, and exploration of, the suburbs. A challenge for many suburban arts administrators, however, is how to reach an interested audience. When a newly hired public programmer for two visual arts institutions in Mississauga had trouble engaging what she characterized as a suburban bedroom community audience, she sought to bolster her event attendance numbers by shuttle-bussing members of the downtown Toronto arts community out to the suburbs. Although she was initially reluctant to organize artistic field trips to suburbia and to mobilize her downtown cultural networks to generate a temporary audience, she quickly realized that where *she* perceived an absence of cultural engagement, *local residents* saw the potential for cultural outreach:

> A reporter from *Mississauga News* said to me last night "It's really great that you brought this busload of people from Toronto to show them that things are happening in Mississauga and that Mississauga is going to be a vibrant cultural hub." I never thought of it that way. I thought that I was bringing my audience in because I couldn't find them here and he was actually seeing it the opposite way around, where we can be injecting audiences from other areas, and that's okay. It just increases public perception of Mississauga as being a space that's open and receptive to artists. Hopefully people would then not leave and go to Toronto to be able to support their careers as artists. (public outreach programmer, Mississauga, 29 August 2007)

Although not stated in this quotation, one significant reason why this newly minted outreach programmer struggled to connect with a suburban audience may be that she herself is not a suburbanite. Like the downtown cultural workers who ride the yellow school bus to explore the suburbs through select art events, she, too, undertakes the reverse commute on public transit and has little lived experience of suburban life.

It is an unfortunate irony that all too often newly hired arts administrators and junior support staff of suburban cultural institutions have not grown up in the suburbs and do not currently live where they work. As the public outreach programmer (Mississauga, 29 August 2007) quoted earlier goes on to explain, in her workplace, "It really is a nine-to-five community, and I think that the gallery has been treated very much like a nine-to-five job where everyone that works there doesn't live in Mississauga." Such suburban disengagement is also clearly apparent in the following quotation from the curator of a North Vancouver gallery (20 June 2007):

> There is a sense of fatigue that sets in because all my staff live in Vancouver. The commute becomes taxing. We really are a Vancouver organization that happens to live in North Vancouver. We all live in Vancouver. We travel here from Vancouver. Our audience travels from Vancouver. Our dialogues are in Vancouver. And we go back home to Vancouver every night. That's our psychic home. We just happen to be over here.

Although this curator acknowledges his gallery's physical location *in* the inner suburb of North Vancouver, neither he nor his staff is intellectually or psychologically rooted here; they are not *of* this suburban place.

Among some suburban cultural administrators there also remain social, cultural, and psychological disconnects, as well as lingering judgmental attitudes about suburbia:

> It's been a real challenge for me because I never lived in Toronto or Mississauga before. When I first arrived, it was really something that was very fresh and it was this enormous hurdle, especially out in the suburbs. I never lived in the suburbs in my life. I had worked in the suburbs before when I used to canvass for Greenpeace when I was 19 years old. I'd get sent out to the suburbs and would come back really horrified by the kind of mentality I would encounter, generally speaking. That was my introduction to suburban culture and I was extremely disappointed. In my mind Mississauga was sort of the armpit of North American culture and decadence. (curator, Mississauga, 30 August 2007)

Despite heading up an important suburban institutional art gallery, this curator is overtly anti-suburbia. In this quotation and in the remainder of the interview he dismisses the suburbs as bland, homogeneous, and unimaginative places that come up short in terms of the quality of life that they afford and the quality of art that they produce. The suburbs for him are a necessary place of work to which he commutes from his downtown art-district home. Any socializing, networking, or exploring in his free time occurs downtown, not in the suburbs.

Projects of (Sub)urban Exploration and Pioneering

The culture of exploration has a long and romantic history rooted in grand nautical and terrestrial voyages of discovery through which the figure of the explorer as pathfinder and trendsetter mythically extends the frontiers of material and imaginative space (Driver, 2001). From ancient Greek travellers onwards, exploration has been variously undertaken to acquire wealth, resources, national prestige, and/or knowledge (Driver, 2001). The history of geographical exploration is not a domestic activity conducted from the pillowed comfort of a living room armchair. The physical act of exploration is intended to take one beyond the confines of the home and the familiar. An explorer is commonly defined as someone who goes into the unknown for reasons of inquiry rather than for mere travel (Fleming & Merullo, 2005). In reality, other than the depths of the ocean, there is very little on Earth that is physically new to be discovered; instead people from one place come to realize and to announce to others that another place exists. Projects of exploration also occur at the frontiers of knowledge and space. Subject to the intellectual curiosity of scholars, journalists, and cultural workers, the suburbs have been positioned since the mid-twentieth century as material and conceptual frontier outposts worthy of exploration.

The frontier has been defined as the territorial margin of jurisdictional influence and civility (Titley, 1999). It is associated with the rationalization and legitimization of the New World in the nineteenth century where frontiersmen and women were described as "intrepid adventurers who brave[d] physical hardship to tame the new land and whose labours pave[d] the way for a material Nirvana to be enjoyed by themselves and their descendents" (Titley, 1999, viii). Pioneers, like explorers, enter difficult spaces from the outside and through innovation and hard work seek to overcome the challenges of starting out in new and unfamiliar places. As an influential myth of imperial discourse,

the frontier motif has transcended historical boundaries and been use-
fully applied by urban scholars to an analysis of cities (e.g., Smith, 1996).
A common theme in the historiography of North American suburbanization is the presentation of "suburbanites as 'pioneers,' moving into
virgin territory and settling down into homes and communities there"
and in so doing "pushing the suburban frontier further into the country" (Clapson, 2003, 145). Women, in particular, actively contributed to
the settlement process whether in the wilderness or the suburb; this was
not an individual process, but a collective effort. This chapter reveals
how cultural workers in the contemporary period can be positioned in
similar ways at "the heart of pioneer neighbourhoods in the new subdivisions" (Clapson, 2003, 146).

Artists have been labelled "pioneers" for their involvement in the
complex process of urban transformation. Sociologist Sharon Zukin's
(1982) influential book *Loft Living: Culture and Capital in Urban Change*
has inspired much research on the relationship between artists and
gentrification, redevelopment, and regeneration (e.g., Cole, 1987;
Podmore, 1998; Foord, 1999; Ley, 1996a & 2003; Phillips, 2004; Slater,
2004 & 2005). Zukin's study examines how artists were attracted in the
1960s to the SoHo district in New York City because of the spatial and
aesthetic potential of the abandoned industrial warehouses as live/work
studios. In the 1980s, many of the artists living in this unique residential
concentration were displaced through real estate speculation and the
widespread appeal of loft living to the middle and upper classes.[1] David
Ley (1996a) has confirmed that the relationship between artists and gentrification that Zukin observed in New York City is apparent in Canadian
cities as well. Ley (1996a, 191) asserts that "the urban artist is commonly
the expeditionary force for inner-city gentrifiers, pacifying new frontiers ahead of the settlement of more mainstream residents." Inspired
by David Ley and Sharon Zukin's work, my own research (Bain, 2003;
2004a; 2004b; 2005; 2006; 2007) has explored the relationship between
artists and cities, with a particular focus on the important role played
by neighbourhoods in the construction of artistic occupational identities. I have portrayed inner-city neighbourhoods – where art is under-
explored, where studio space is affordable, where grassroots ventures
are possible, and where decay and poverty are visible on the streets – as
the untamed, uncharted urban cultural frontiers where worlds of creative possibility await the adventurous artist. In these inner-city contexts
the frontier motif resonates with the identity narratives of contemporary
urban artists for its celebration of the values of rugged individualism,

independence, self-sufficiency, and freedom through hard work and self-sacrifice (Furniss, 1999). But what of suburban cultural workers? If the suburb has been so comfortably framed as an urban frontier, why then are cultural workers rarely represented in the popular and scholarly imaginations as suburban pioneers?

Suburbia, especially in the immediate postwar period, "set the standard of what was normal in society – the nuclear family – it also became the standard against which what was abnormal could be judged. Those who fell outside the rubric of the nuclear family and for physical or moral reasons could not be accommodated by suburbia were gays, single parents, the poor, artists and intellectuals" (Sowden, 1994, 87). A legacy of this standard has carried over to the present day, positioning artists as socially out of place in the suburbs. Instead, the artistic labour and creative practice of cultural workers are most comfortably associated, David Ley (1996a) argues, with the energy and intensity of the central city. In addition to being excluded from the nuclear family rubric, artists, maintains Ley (1996a, 206), have themselves rejected the suburbs as "too standardized, too homogeneous, too bland, too conformist, too hierarchical, too conservative, too patriarchal, [and] too straight" in favour of "authentic places" in the central city that are "marginal to conventional middle-class definitions." My research on artistic practice in Toronto has also shown that in the artistic and public imaginations, cultural production is most closely allied with the "authenticity" of downtown art districts and studio buildings in converted industrial warehouses.

Opinions differ on how to define authenticity. The term authenticity is often used in the context of heritage preservation to refer to the use of reconstruction and renovation to return a building to its original state. Thus a building or a landscape would be interpreted as authentic if it accurately reflects an earlier, original form or representation. For artists, authenticity often refers to original architectural character and the rough tangibility of decaying buildings and objects in the urban landscape, which provide artists with a sense of historical lineage and creative possibility. For a cultural product and a cultural worker to be considered authentic, legitimate, and of value, they must be geographically positioned as close to the urban centres of art world power as possible (despite the rapid gentrification of these landscapes and the loss of their original purpose and residents). If suburbs are deemed to have limited cultural value – and what little there is can be difficult to identify because it is so spatially dispersed – then there would appear to be no incentive to value the activities of suburban cultural workers as pioneering. We are,

however, witnessing a scholarly shift away from the denigration of the
suburbs to a cultural valuation of their many complexities (McAuliffe,
1994). Perhaps the cultural worker no longer cuts such an unusual
figure in suburbia.

In his book *Suburban Landscapes*, Paul Mattingly (2000) paints a detailed
historical portrait of Leonia, New Jersey, a suburban community outside
New York City with a thriving artist colony. From the late nineteenth
century onwards, generations of artists "reflected a range of values recognizable to suburbia (such as the primacy of ordinary routine, adaptation to often inhospitable terrain, and the supportive force of traditional
family roles), but they also actively promoted and reshaped their suburb's communal culture, especially the reconstruction of their historical
inheritance" (Mattingly, 2000, 7). Throughout Western history we find
other examples of suburban artistic colonies. Elizabeth Wilson (2000),
in her historical portrait of Bohemia and bohemians, for example, writes
about Montmarte outside the boundary of Paris, and Schwabing, a suburb of Munich, the artistic capital of Germany, in the nineteenth century. Montmarte, with its lower wine prices, became a renowned artistic
countercultural haunt and an entertainment district of cafés, bars, caba-
rets, and working-class music halls. Schwabing was a suburban residential
neighbourhood that became a "hotbed of cultural activity" around 1900
because of "the sheer number of intellectuals, political figures, publishers, and artists who lived and worked" there (Wilson, 2000, 34). Into the
twenty-first century, cultural workers continue to engage in pioneering,
exploratory, and interventionist acts in suburbia. The historic and ongoing presence of these cultural workers in the suburbs demonstrates that
suburbia has not eroded the values of individualism, innovation, and
originality so cherished by the artistic avant-garde (Sowden, 1994). The
remainder of this chapter illustrates that the suburbs can be a place of
contradiction and flux rather than stability (Webster, 2000).

A Gendered Suburban Landscape

Traditional historical scholarly accounts of the development of
nineteenth-century middle-class residential suburbs portray them as a
paradise for men – a place that provided an escape from the congestion,
pollution, and corruption of the city and a necessary separation of work
life from family life (Whitehand & Carr, 2001). A spatial and gendered
division of labour between the public spaces of the city and the private
spaces of the home helped to protect the aspirations of opportunistic

middle-class Victorians for order and family stability. As Mona Domosh and Joni Seager (2001), among other scholars (e.g., Silverstone, 1997; Cunningham, 2000; Head, 2000; Oswell, 2002; Clapson, 2003; Stevenson, 2003), have documented, cities were characterized as masculine, public, egalitarian, extroverted, spectacular, and active. Suburbs, by contrast, were characterized as feminine, private, deferential, introverted, emotional, and passive. In such a polarized conceptual framework suburbs signified "domesticity, repose, closeness to nature, lack of seriousness, mindlessness, and safety" (Susan Saegart cited in Clapson, 2003, 125). Such simplistic and reductionist signification has carried remarkable potency.

Women and the suburbs continue to be intimately connected in a gendered myth (Burns, 2011) that equates the suburbanization of culture with the feminization of culture (Silverstone, 1997).[2] Ironically, however, when Statistics Canada labour force data on employment in professional occupations in art and culture from 1996 to 2006 are examined, a gender shift can be seen in the increased percentages of men living and working in the suburbs. Figures 3.2 and 3.3 document an increase in suburban cultural workers, particularly men, for the case study areas on the peripheries of Toronto and Vancouver. In Figure 3.4, the percentage change of cultural workers by gender between 1996 and 2006 can be seen and the number of male suburban cultural workers has increased more rapidly than the number of female cultural workers.

Suburbia has come to epitomize the intersection of the commercial and the domestic realms (Head, 2000). Within the art world, in particular, the suburbs are portrayed as "a world of manufactured tastes, of pristine domesticity and social conformity, all of which is [supposedly] antithetical to the artistic tradition" (Sowden, 1994, 78). As I have written elsewhere in articles on artistic identity construction, myths and stereotypes play an important role in helping to create an artistic identity and to project it to others (Bain, 2005). Such artistic occupational mythology is rooted in reverence for the heroic male genius who seeks spatial and social isolation in order to remain committed to his creative vision. It celebrates the creativity, originality, and imagination of individual male artists as expressed through masterpieces, and it glamorizes the image of the starving bohemian rebel living in a central city garret or post-industrial loft studio sacrificing status, money, and material comfort for the supposed freedom to pursue creative expression (Bain, 2007). The suburbs do not fit comfortably within this masculinized mythological framework that symbolically locates creativity in the downtown core

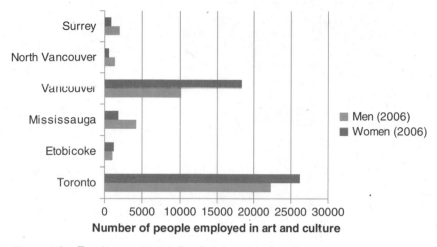

Figure 3.2   Employment in professional occupations in art and culture in 2006 by gender (Statistics Canada, 2006).

of cities. The "coupling of suburbia with femininity" and with social compromise means that some cultural workers perceive the supposedly "female-dominated suburbs" as "a threat to creativity" (Sowden, 1994, 83) and seek to distance themselves from them.

Neither Lost nor Out of Place

The sociologist Richard Lloyd (2006, 8) in his book *Neo-Bohemia: Art and Commerce in the Postindustrial City* documents how one working-class Chicago neighbourhood marked by deindustrialization and population decline was transformed over several decades. Wicker Park went from relative obscurity into a staging ground for young avant-garde artists and "a celebrated center of hip urban culture." The engine of urban redevelopment in this neighbourhood, north and west of the downtown Loop, was a critical mass of artists who constituted the local art scene, combined with the purchasing power of educated young professionals employed in new media and design firms, who supported the local entertainment economy by frequenting local galleries, performance venues, bars, restaurants, coffee shops, and boutiques.

Lloyd (2006, 11) developed the term "neo-bohemia" as "shorthand for both a distinctive sort of urban district and an associated style of

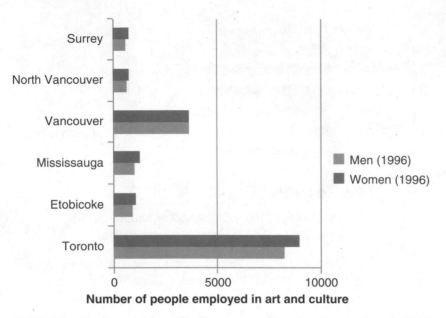

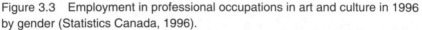

Number of people employed in art and culture

Figure 3.3   Employment in professional occupations in art and culture in 1996 by gender (Statistics Canada, 1996).

life," mythically attributed to the artists' quarter of nineteenth-century Paris and recreated in twentieth-century Chicago. In nineteenth-century Paris, the term "bohemian" was used synonymously with the words "beggar" and "gypsy" as a pejorative to describe self-chosen social outcasts who sacrificed money, security, and social position for the freedom to pursue creative practice that such marginality afforded. In the early twentieth-century North American context, bohemia was associated with the countercultural movements that opposed the rationalized organization of labour and commerce under industrial capitalism. Bohemia materialized in artistic districts such as New York City's Greenwich Village and East Village, Los Angeles's Venice Beach, and San Francisco's North Beach and Haight-Ashbury. The new bohemia of the late twentieth century, however, is not as marginal to the urban economy or to society. Lloyd (2006, 69) argues that "new bohemian districts are increasingly common features of the contemporary urban landscape" because they have a valuable role to play in enhancing property speculation and post-industrial capitalist enterprise. New bohemian districts, like Wicker

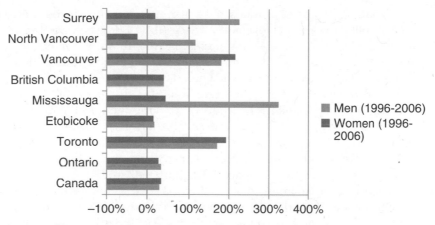

Figure 3.4   Percentage change in employment in professional occupations in art and culture by gender 1996–2006 (Statistics Canada 1996 and 2006).

Park in Chicago, with its low-rise, derelict factories and warehouses, and homes of immigrant labourers, are characterized by Lloyd as eclectic, diverse, intense, and authentic. Through interviews with local cultural workers, Lloyd documents how artists who live and work in Wicker Park often celebrate and identify with the urban grittiness, edginess, and instability of their neighbourhood. Such spatial identification with the neighbourhood is then worn as a badge of cultural distinction that proclaims commitment to the artistic profession. In Lloyd's (2006, 75) estimation, artists are "committed urbanites" who "fold the representation of neighbourhood decay into their picture of authentic urbanism, even as their presence contributes to the reversal of many of its effects." He concludes that despite "the staggering rise of suburbs and edge cities as sites of residence and economic activity, cultural production still privileges the old center city as a generative milieu and site of fantasy" (Lloyd, 2006, 244). This chapter takes Lloyd's conclusions as its starting point. It considers whether artists can be committed suburbanites and whether neo-bohemia can exist in a suburban cul-de-sac.

The density, distinction, and relative ease of access to cultural infrastructure in the central core of North American cities have an almost magnetic pull, drawing artists to live, work, and help create art districts out of inner-city neighbourhoods. Take, for example, the Scarborough-based installation and performance artist Erika DeFreitas, who lives in an eastern inner suburb of Toronto. In a playful effort at urban beautification, DeFreitas initiated her *Something Pretty Cozy* project (2003) along Toronto's main downtown art axis, Queen Street West, before moving it on to Brooklyn and then San Francisco. She crocheted bright and cheerful covers for public street fixtures: parking meters, bollards, fire hydrants, and pay phone handsets (www.erikadefreitas.com). Her covert textile street art, sometimes referred to as "knit graffiti" or "yarn bombing," is part of an international guerilla art movement whose practitioners seek to transform public space with "donated" works of art created out of yarn (Moore & Prain, 2009). When asked at a community lecture in 2008 on the "Suburban Art Scene" sponsored by the Scarborough Arts Council why she had chosen to intervene in the urban rather than suburban landscape, she explained that her interventions get lost in suburbia; suburban residents do not seem to notice, discuss, or respond to her work.

Suburbia can be interpreted as a place of distance and journeys, where public transit, shopping centres, and public libraries become the civic living rooms and the bus stop, the bus, the commuter train, and the mall are places of encounter on the way to somewhere else. In Doreen Massey's (2005, 130) words, to "travel between places is to move between collections of trajectories and to reinsert yourself in the ones to which you relate." The act of relating can occur in different locales: for some it is in an office, for others it is on a bus or in a shopping mall, or in a community centre or a home. But suburban bodies in motion between work, recreation, shopping, and home must eventually lose momentum, slow down, and pause to rest. The traditional site of respite and social reproduction in suburbia is the single-family home in a residential subdivision. As the work of Gordon Matta-Clark (1943–78) nicely illustrates, the suburban home can be both a site for cultural production and an artistic object.

Destructive Impulses within the Suburban Home

Pamela Lee has provocatively titled her retrospective of the oeuvre of American artist Gordon Matta-Clark *Object to Be Destroyed* (2000). Trained as an architect and active in the New York City SoHo art scene in the 1970s, Matta-Clark was best known for his "'building cuts': sculptural transformations of abandoned buildings produced by cutting and dismantling a given architectural site" (Lee, 2000, xii). In an era of conceptual art that privileged the idea of the art object and the process over the finished product, Matta-Clark's art-making played out through the dissection of architectural space and was intimately tied to the destruction of the original work.

In 1974 in Englewood, New Jersey, a bedroom community for commuters to New York City, Matta-Clark took several months to transform an abandoned, two-storey, clapboard, single-family suburban home into a work of art titled *Splitting*. His dealers had bought the soon-to-be-demolished house as a speculative real estate venture and granted the artist the use of the building. He began by clearing out all of the possessions left behind by the previous tenants and storing the biographical detritus in the basement. Then, using a chainsaw as his dominant sculptural tool, he cut two parallel vertical lines in the middle of the house and scored the walls, floors, stairs, and ceilings. He chiselled away the cinderblocks reinforcing the foundation and, using building jacks, tipped the house back on itself. Material from the middle of the house was removed, along with the eaves, to produce a sectioned view for a general audience (Walker, 2009).

Friends and colleagues were bussed from the art world epicentre in Manhattan "like schoolchildren on a class field trip [to] explore the suburban exotica" (Lee, 2000, 28). A conventionally private space had been rendered permeable to the public gaze. Visitors remarked on the play of light and shadow and "the precarious balance the work str[uck] between exterior and interior, urban and suburban, public and private, violence and enlightenment" (Lee, 2000, x). The cleaved suburban house has been variously interpreted by art critics as: a house and nation divided by class, ethnicity, and race; a violent rape of the feminine-coded domestic sphere; and a challenge to the middle-class dream of the suburban home steeped in the ideologies of privacy, autonomy, and stability (Lee, 2000).[3]

In the tradition of Gordon Matta-Clark, the Arbour Lake Sghool investigates the suburban home as unfinished art object and meticulously documents on its website the activities that occur within it. They "ironize the space of the house, traditionally regarded as a cipher of privacy ... stability, permanence, security" (Lee, 2000, 2 & 21). They, too, seek to develop an artistic process that enacts "the ultimate freedom of private ownership, possession through destruction" (Lavin cited in Walker, 2009, 123). Differences can be found in the degree of occupation and the extent or permanence of the destructive impulse. Matta-Clark dissected abandoned or condemned buildings that he neither owned nor resided in. The Arbour Lake Sghool has chosen instead to transform their domestic life and its material trappings in the Calgary outer suburb of Arbour Lake into art. As was the case with the English family who transformed a bedroom in its council flat on the outskirts of Liverpool into the Institute for the Art and Practice of Dissent at Home (www .twoaddthree.org), the home can become a site of disobedient activism rather than a place of consumption.[4]

Neo-Bohemia in a Residential Cul-de-Sac: The Arbour Lake Sghool

Calgary, Alberta is Canada's fastest-expanding metropolitan region and has the largest footprint of any North American metropolis. With a 2006 population of approximately 1 million living in an area of 850 square kilometres, Calgary is a city overwhelmed by its suburbs. In the north-western suburb of Arbour Lake, beside the international airport and the Trans-Canada highway to Banff, a five-member art collective called the Arbour Lake Sghool was founded in 2003. This creatively deviant group of artists, musicians, and tradespeople has a mandate to "provide a stage for the creation and display of artistic or critical projects in a way which explores and engages our suburban setting" (www.thearbourlakesghool .com). Despite this art collective's interventionist mandate of community engagement, it is not on the community radar of the Arbour Lake Residents Association, whose website makes no mention of the work of the Arbour Lake Sghool or any other local artists.

The five resident male artists of the Arbour Lake Sghool collective have re-socialized themselves into the suburban home – a space traditionally presented in public policy and media images as feminine. This group of young men live under one roof and identify as a self-chosen family. They share a joint-signature bank account, the cost of rent, and the domestic responsibilities of shopping, cooking, and cleaning, and they collaborate on the creation of art. The artists in this friendship circle have various paying jobs and the money they earn is pooled and used to collectively finance their artistic practice within an economy of generosity and voluntary reciprocity.

Members of the Arbour Lake Sghool collective have used art to reconfigure the left-over space of the neighbourhood cul-de-sac into a communal public space of collective belonging, social exchange, and informal encounter (Tonkiss, 2005). While some neighbours ask questions of the artists, by way of engagement with their work, not all local residents have been as enquiring or as receptive. Some remain indifferent, hostile, or non-communicative, frequently using property by-laws as a way to enforce what they regard as socially acceptable codes of behaviour. The front yard of the Arbour Lake Sghool, a site of numerous artistic interventions that challenge understandings of landscape design and property ornamentation, has come under significant scrutiny from neighbours (Figure 3.5).

In an exhibition catalogue essay the Toronto journalist and art critic Gary Michael Dault (2001, 5) asks: "What kind of art might come from artists living in the suburbs?" His flippant reply does little to answer the question: "I have no idea. No idea and too many ideas, actually. The subject is always, at least initially, dispiriting." The creative labour of the Arbour Lake Sghool is not remotely dispiriting. With a predominantly unfenced front yard on a highly visible corner lot, the Arbour Lake Sghool has a useful venue for showcasing outdoor ephemeral works of art – everything from a geodome constructed out of discarded pizza boxes, to a grass Mohawk, or a large two-storey, twenty-person hammock.

The Arbour Lake Sghool treats its works of art not as objects to be displayed in a gallery but more as ephemeral processes of creation. Its cross-disciplinary, experimental art-making could be described as "post-studio" "social practice" (Jackson, 2011, 1). For Shannon Jackson (2011, 12), art as social practice brings together aesthetics and politics; it "is a term for art events that are inter-relational, embodied, and durational." The artistic labour of the Arbour Lake Sghool fits within a relational turn in contemporary art that was initiated in the last decade of the twentieth century (Purves, 2005). French art critic Nicolas Bourriaud (2002) has developed the phrase "relational aesthetics" to describe the work of artists, like the Arbour Lake Sghool, who use "inter-subjective exchange" (the sociability of human interaction and encounter) as their creative "substrate." According to Bourriaud (as cited in Fowle & Larsen, 2005, 19), relational artists find gaps within capitalism to create "'new social models' in the public sphere." For the Arbour Lake Sghool, the events, the activities, the online documentation, the audience responses are all details that form and inform the works of art. The Sghool's suburban home is a continuous, lived archival art project altered daily, much like Kurt Schwitters's "merzbau" – an installation of everyday objects built out of his home studio in Hanover, Germany between 1919 and 1937.[5]

The Arbour Lake Sghool has transformed the suburban home into a dynamic site of cultural production. Spaces both inside and outside their residence have been redesigned as expressions of personal taste and identity. The suburban cul-de-sac and the suburban yard have

Figure 3.5   Creative detritus in the Arbour Lake Sghool yard and driveway stands in contrast to the landscaped order of the neighbouring lawn (photograph by A. Bain).

become sites of protest and resistance to social norms and property rules through artful play and creativity. The freedom to play is a valuable form of creative exploration, where accidents, failures, and inspiration can mix to foster cultural dialogue. But such temporary and playful creative transformations are not without social conflict. Artistic interventions invariably make demands of people; they have the potential to produce "unpredictable dissonances" and to "generate reverberations that do matter" (Mahtani, 2002, 436). The social tensions wrought by conflicting values and ideals are a lived reality for those who make art in suburbia. This collective of artists embraces such tension, and the media coverage it generates, as a vehicle for dialogue, through their art and on their website, about the possibilities of suburbia. (For more details on the Arbour Lake Sghool please see Appendix 1.)

Conclusion: A New Generation of Suburban Bohemians

The creative impulses of the Arbour Lake Sghool fuse labour and product, work and play. Like the nineteenth- and twentieth-century bohemians before them, this collective of artists is motivated by a "disobedient spirit which seeks to dissect the world it lives in" and to reimagine an alternative (www.thearbourlakesghool.com). The opening quotation on the collective's website is from work on civil disobedience by the anarchist Henry David Thoreau: "The greater part of what my neighbors call good I believe in my soul to be bad and, if I repent of anything, it is very likely to be my good behavior." As Elizabeth Wilson (2000) documents in her book *Bohemians: The Glamorous Outcasts*, bohemian life has been constructed as a kind of carnival in which conventional boundaries are transgressed and the repressive authority of bourgeois society defied through excess, disorder, and humour. Wilson shows how bohemia emerged through literature and painting as an oppositional counter-space (a destination) and an unconventional way of life (a journey) created by artists, writers, intellectuals, and radicals contrary to the dominant culture. By living and behaving differently from mainstream society, in ways sometimes deliberately calculated to shock, bohemians dramatized their differences and sought to merge the aesthetics of art into the politics of life.

Practising a bohemian lifestyle can stereotypically involve, as van Heur (2009, 1546) notes, "laziness, unfocused attention and free time" as well as disdain for conformity and upward mobility. Such qualities Robert Beauregard (2006, 135) deems as "incongruous with the suburbs" and "against the grain of postwar prosperity and mass consumption." Yet a new generation of suburban bohemians and explorers, as exemplified in this chapter by the Arbour Lake Sghool, is explicitly using suburbia as its imaginative foil and demonstrating that the suburbs can be creative. There has been "an outpouring of creative works of people who grew up, or currently live, in the suburbs seeking to explore their suburban experiences and memories through their art" (Stevenson, 2003, 128). This new generation has created Lefebvrian counter-spaces that are "against quantity and homogeneity, against power and the arrogance of power, against the endless expansion of the 'private' and of industrial profitability; and against specialized spaces and a narrow localization of function" (Lefebvre, 1991, 382). Such counter-spaces are part of an alternative spatial system of different arrangements, practices, and norms that run against the grain of society. These sites explicitly challenge the dominant organization of space through creative reinvention and adaptive reuse. Counter-spaces celebrate use value over exchange value[6] and, in so doing, refuse to succumb to the predatory logic and bureaucratic management of capital (Tonkiss, 2005). As is discussed in chapter 6, such counter-spaces of suburbia have the potential to augment the place complexity of the suburbs.

Chapter Four

Why (Not) Here? Valuing the Intermediate Landscape

I believe things should keep changing. An artist's life is not static and where you live should be the same.

—(painter Charlotte Ardizzone quoted by Bamber, 1997)

The quotation that opens this chapter draws a deliberate parallel between the life of an artist and the trajectory of a suburban landscape. Each seems to suit the other. Much like the lives, careers, and creative outputs of cultural workers, the imaginative and material landscapes of suburbs are in a constant state of transformation and emergence. Suburbia appears as "evolutionary" and "transitional" in its physicalities and lived realities as the cultural workers who reside there (Bourne, 1996). Suburbia, as Judy Giles (2004, 30) has shown, is not just a geographical response to the radical alteration of living arrangements brought about by industrialization and urbanization; it needs to be appreciated "as an architectural and social space, an ideal, a producer of cultural meanings, and a psychic and emotional landscape." The suburb as interpreted by contemporary cultural workers is often appreciated and negotiated as a series of juxtapositions rather than as a synthetic whole.

To better understand the lived complexity of suburbia, the fundamental question that is posed and answered in this chapter is: why here?[1] From this basic question three interrelated questions extend. Why have cultural workers chosen to live and work in the suburbs? What do they perceive as the main advantages and disadvantages to the production of culture and the practice of creative labour in a suburban location? And in what ways do the suburbs meet or not meet the daily creative needs of cultural workers? A consideration of these questions requires a preliminary theoretical return to David Ley's concept of the "new middle class" and Richard Florida's concept of the "creative class," from which two new questions emerge. Are cultural workers members of a distinctive socio-economic group with unique locational aspirations? And do cultural workers follow rather than lead the middle class in their residential choices?

From the "New Middle Class" to the "Creative Class"

In a now-classic book, *The New Middle Class and the Remaking of the Central City*, David Ley (1996a) draws examples from Toronto, Montreal, Vancouver, and Ottawa to examine the spatial involvement of urban artists in the gentrification of the Canadian inner city. By using occupational locational quotients as a measure of geographical concentration, Ley demonstrates that between 1971 and 1991 the numbers of artists residing in the central city increased for Canada's three largest CMAs and decreased for the surrounding suburbs. He shows that over these three decades, bands of location quotients for artists decreased with increasing distance from the central city. With correlated census data, Ley (1996a, 194) asserts that in 1971 artists had similar locational characteristics to. middle-class households with homes in the central areas of Canada's largest cities. The inner-city neighbourhoods that had marked concentrations of artists were characterized by: above average social status; proximity to élite districts, universities, and hospitals; few blue-collar workers; a greater than average number of students; and an Anglo-Canadian ethnic orientation. In addition to "being bound to the middle class, artists are also its advancing or colonizing arm, invading new realms of convention, style and consciousness" (Ley, 1996a, 191). In the idealized stage model of gentrification it is artists and cultural professionals who often arrive first in a neighbourhood followed by professionals in education, health care, welfare, and the sciences, and finally by private-sector employees in management, sales, and finance. By 1991, Ley determined that consecutive waves of gentrification had secured the relationship between artists and the new middle class.[2] For Ley (1996a, 191), urban artists are connected to the new middle class in the sense that they lead inner-city property reinvestment for them by "pacify[ing] new frontiers ahead of the settlement of more mainstream residents." Artists value central-city locations, Ley (1996a) argues, for the functional advantages they provide of affordable studio space as well as links to customers, suppliers, and the downtown art scene. In Ley's (1996a, 195–6) estimation, "artists are quintessential city people" who favour downtown locales because they "provide oppositional spaces, social and cultural distance from the everyday life of the suburbs, its conformity, standardization, and understimulation." The suburbs in this framework hold negative value; to artists and middle-class gentrifiers they have a stigmatized and "spoiled identity" (Ley, 1996a, 205). This sentiment is neatly captured in a quotation from a Vancouver sculptor whom Ley (2003, 2534) interviews in a later article that furthers his argument about the central city as "artistic habitus": "Artists need authentic locations. You know artists hate the suburbs. They're too confining." The story that Ley tells about artist-led gentrification and the keen spatial differentiating sensitivities of artists is an important one, but it neglects and, with the strong judgments passed about the suburbs, further marginalizes the lived experiences of suburban artists.

This book shows, however, that a suburban life has become a necessary reality for many contemporary creatives. While suburbia continues to present some obstacles to professional cultural production, suburban cultural workers find much to value in the suburbs. This chapter shows that in their shared valuation of suburban spaces and amenities, suburban cultural workers have actually retained strong ties to the "old" middle class and asserts that the locational preferences of suburban cultural workers are not all that different from those of the traditional middle class.

The relationship of artists and cultural workers to the middle class is an interesting one. David Ley (1996a, 189) describes this relationship as symbiotic: "The artist has a distinctive bridging function in conveying novel and sometimes iconoclastic styles to the middle class from which he and she are commonly drawn. The artist serves a social role as a broker of fashionable middle-class taste, demarcating the new frontiers of cultural distinction." Drawing on Ley's work, I have discussed elsewhere (Bain, 2005; 2006) how cultural workers are often raised within middle-class families, educated with middle-class young people, and grow up to socialize with the middle class and to service them with their cultural products. Through their creative labour, cultural workers mediate between the occupational categories of trades and professions. In her book *Thinking in Place: Art, Action and Cultural Production* Carol Becker (2009, 36) explains that art-making, in addition to "being understood as a service, also has been considered a craft, so it is often not understood to be in the same league as other serious intellectual work. Craft-making, after all, is something that can get you dirty – not unlike, in fact very much like, manual labour." Working with one's body for a precarious and irregular salary might be considered working-class, but the occupational title "professional" in combination with the possession of cultural capital (e.g., art education, "aesthetic disposition" [Bourdieu, 1993], hipsterness, progressive ideas, and place in the contemporary arts scene) can render cultural workers members of society's elite. In cities, neighbourhoods, and cultural institutions, the rebelliousness of the art world, as embodied by the artist, "collides with the actual and desired economic power, aspirations, and conventionality" of the ruling, moneyed classes (Becker, 2009, 38). In such class collisions cultural workers are often "conceived of as middle-class people slumming and playing poor" (Solnit & Schwarzenberg, 2000, 18). Neither completely accepted as middle-class nor as working-class, cultural workers appear to hold a transient economic status that Richard Florida deems to require special designation.

Richard Florida's (2002; 2005) "creative class" is a diverse group of professionals across the occupational spectrum who are paid to think and problem solve. They include lawyers, bankers, doctors, scientists, engineers, architects, designers, and business entrepreneurs, who are said to share approaches to problem-solving and work practices that value creativity, individuality, diversity, and merit. At the core of his socially constructed class are cultural workers. Like their creative class colleagues (although with a significantly lower average annual income), cultural workers are said to want to locate in urban places that are trendy and cool and that provide sufficient stimulation, diversity, and depth of experience to inspire creativity.

Among a range of criticisms raised of Florida's work are questions about whether there can be a common class identity for a class whose membership is said to range from bohemians to executives (Markusen, 2006; Indergaard, 2009; Krätke, 2010). Markusen (2006, 1921) argues that the creative class is a "fuzzy" concept that groups occupations together that "exhibit distinctive spatial and political proclivities" simply "on the basis of educational attainment and with little demonstrable relationship to creativity." Members of the creative class do not share "class interest, outlook, or social patterning of behaviours" (Markusen, 2006, 1924). Krätke (2010) concurs, asserting that the creative class is merely an arbitrary aggregation of unrelated occupational groups based on co-location. Members of this idealized class are said to be attracted to gentrifying central-city neighbourhoods and to a particular form of urbanity that exacerbates inequalities (Peck, 2005; Wilson & Keil, 2008; Evans, 2009; Catungal & Leslie, 2009b).

Florida's creative-class concept celebrates the creative talents and practices of cultural workers as distinctive within the labour market and as significant to the health of the new economy of cities. Ironically, however, as this chapter shows, the place attributes that attract cultural workers to the suburbs are not all that different from those that have attracted the middle classes through history. Cultural workers may be the headliners of Florida's creative class, but they can also be quite conventional in their residential locational choices.

Why Here?

Numerous Canadian cultural workers, motivated by both necessity and choice, live and work in the suburbs, negotiating popular (mis)conceptions of suburbia as both "suburban dream" and "suburban nightmare" (Hall & Lee, 2010). The suburban dream perspective sees "suburbia as a symbol of goal-fulfillment: the place where families can buy a home, avoid urban ills, commune with nature (and each other), send their children to good schools, and climb the social ladder" (Hall & Lee, 2010, 3). Conversely, the suburban nightmare perspective dismisses suburbia as a "cultural desert" of commuting and consuming, a place devoid of "diversity, cosmopolitanism, political culture, and public life" (Sharpe & Wallock, 1997, 310), and a place where creativity cannot possibly establish sufficiently viable roots to flourish. In very general terms, the academic and popular discourse on suburbia is divided "between those that support the idea of the suburbs and the quality of life they offer, and those that endorse the view that suburbs are 'non-places' where nothing exciting ever happens and no one interesting, different or creative lives" (Stevenson, 2003, 123–4). This tension between the positive and negative characterizations of suburbia parallels the advantages and disadvantages that cultural workers themselves associate with their decision to live and work in the suburbs of Canadian cities.

Cultural workers are both pushed and pulled to the fringes of Canadian cities. In conversational, semi-structured interviews, my sample of cultural workers from the inner and outer suburbs of Toronto and Vancouver were asked their reasons for living and working in the suburbs and what they perceived as the main advantages and disadvantages of such a location. An analysis of their answers reveals a mosaic of reasons as to why cultural workers locate in suburbia. Figure 4.1 summarizes

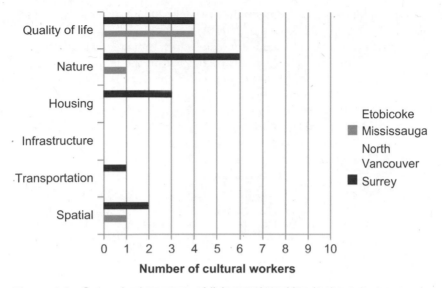

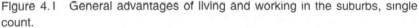

Figure 4.1 General advantages of living and working in the suburbs, single count.

the main spatial, transportation, infrastructural, housing, environmental, and quality-of-life advantages identified by cultural workers, and indicates the number of informants who mentioned each one. Figure 4.2 provides more detail, quantifying the particular advantages simultaneously identified by informants.[3]

Bucolic Suburban Fantasies within Industrial Realities

Quantifiably, the most favourable feature of the suburbs identified by cultural workers (particularly those residing in inner suburban locations) is proximity to nature. The idea that natural environments can foster cultural and intellectual creativity has a long history within the arts. Seasonal or periodic relocation to a more natural setting for inspiration – an artist's retreat – is certainly a common phenomenon in the visual, musical, and literary arts.[4]

Many of the cultural workers interviewed for this study communicate a romanticized understanding of suburbs as places that are close to nature and the countryside, and are more rural than urban. Suburbs have been

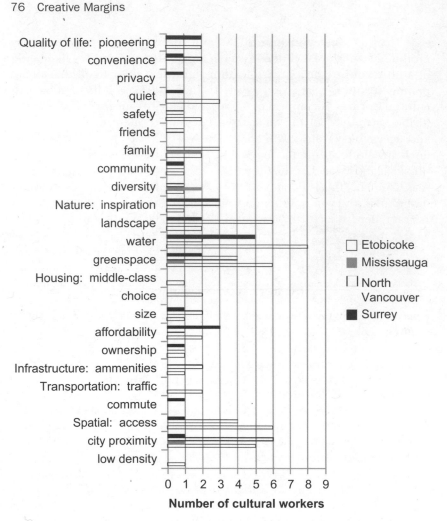

Figure 4.2   Advantages of living and working in the suburbs, multiple count.

interpreted by Bourne (1996, 179) as a product of "rural nostalgia" that reflects "a desire to return to the countryside and rural roots, but without also severing connections to the urban core." Such nostalgia is evident in the following quotations: "I really like the fact that here you're so close to, and almost completely surrounded by, water, and … in less then 10 minutes I'm in the country. I really like the access to nature that way" (painter, Surrey, 22 June 2007); and "Because we're surrounded by mountains and forest, it doesn't seem as suburban to me. It's like being in a little country or a little town but right next door to all that's happening" (potter, North Vancouver, 15 June 2007). For these suburban cultural workers, the appeal of Surrey and North Vancouver is the "rural idyllic *appearance*" of the landscape that renders nature visibly present and accessible (Watt, 2009, 2882). In North Vancouver, this potter is far from alone in identifying the appeal of the natural beauty of the physical landscape. For the director of the North Vancouver Community Arts Council, "this is paradise … you've got the beaches, you've got the mountains, and you're fifteen minutes from downtown" (29 May 2007). A storyteller from North Vancouver (21 June 2007) elucidates:

> Right away we knew that there was a real strong support in the North Shore for the arts. There's a lot of incredible artists here – visual arts, music, drama – because people want to be near the city but also there's such an element of the natural and the beauty and the landscape. People choose to be up here for very specific reasons and not because the land is cheaper or anything like that.

The location choices of this storyteller are echoed in this comment from a musician and colleague (North Vancouver, 21 June 2007):

> I came here to get to a place with more greenery and a little more space in my home … I had a friend who was living in this neighbourhood but she's moved away since. That was what I was drawn to … I spend time out walking in the woods and on the beach. If there's anything that would be an influence on my work, it's probably that … that's a wonderful thing about North Vancouver is that we are never far from a forest or a mountain or water.

The importance of being close to nature is reiterated by a local photographer (22 May 2007) who treats his natural surroundings as his muse: "I'm close by the mountains, so I can get out of the urban environment and move into more appropriate environments and become inspired and connect with the land." In interviews, suburban cultural workers repeatedly highlight the importance of living and working in close proximity to bodies of water, to distinctive landscape features (e.g., mountains, beaches, and wetlands), to parkland, and to open space because it allows them to connect with the natural world around them.

Closeness to nature is an attribute as sought after in the borderlands of the 1820s as in the present day (Hayden, 2003).

The freedom to explore the natural realm on foot or by bicycle, without permission or financial expense, holds appeal for cultural workers. They may imagine that these green spaces are wild natural landscapes far away from the city and its dominant forces of consumption and production, rather than places where nature is managed and manicured. As one photographer from Etobicoke (8 November 2006) explains:

> Those areas below the Lakeshore, I find them calming. The lake is down there. I bicycle a lot. I like the idea of being able to get out of that urban core to just relax and decompress ... The biggest thing that I like around here is the lake and Colonel Sam Smith Park. I love the fact that there's hardly anybody in there. You can go there on a Sunday afternoon or morning and you're just flying around and you'd think you're a long ways from a city core. The Etobicoke Creek, you go in there and you feel like you're 200 kilometres away from Toronto.

It is apparent in this quotation that suburbs can be understood in quite a traditional sense as places that can dispel tensions and produce spaces and times of calm in the midst of the chaos of urban living.

Cultural workers emphasize the importance of finding quiet and solitude in the outdoors, moments of escape from the tensions, intensities, and frenetic pace of urban life and the competing demands of creative labour. Natural environments offer an allure of individual freedom and spontaneity and are a catalyst for individual creativity:

> I live on a road that ends at the creek, sort of a dead end street, and there's an enormous two-acre park tucked away. You can't even see it from the street. You have to walk down and you see that there's this beautiful park. I look at that out the window of my studio and you know it's very much the consciousness that I embrace. It's what I want to see. I don't want to see buildings and I don't want to see traffic, although increasingly I am seeing those too. (composer, Etobicoke, 9 November 2006)

A similar sense of inspiration and escape is expressed in the comments of another musician who lives further south in Etobicoke, just a block away from Lake Ontario:

> The peace and quiet of the area expresses itself in my music ... what it does is it cleanses your brain and allows the muse to happen. There's just

Figure 4.3   The garden that inspires a Surrey painter in her home studio (photograph by A. Bain).

something very therapeutic here. If this were the morning, I'd say, "You want to go into the midst of nowhere? I can get you there in a 10-minute walk." It's incredible the unexplored territory down here (musician, Etobicoke, 24 October 2006).

Many arts practitioners emphasize the significance of local natural landscapes to extend their imaginative capacities and provide inspiration. In the words of another Surrey resident (18 May 2007), who has a home painting studio and is an avid gardener (Figure 4.3), "I realize nature is my spirituality. It's the core of it. It keeps me centred. It's my inspiration. It shows me persistence."

For one printmaker (Etobicoke, 25 October 2006), who has spent the last several years working on a botanical series, the plants she sees cycling and walking through the local parks are an aesthetic prompt: "I'm two blocks from the lake and I love walking down there. When I can, I ride my bike ... the lake, the water, the clouds and the plant life influence my artwork, which I don't think I would have been inspired by, if I wasn't so close to it." Water is a particularly influential feature of the natural landscape that is present in the paintings, prints, plays, and music produced by local artists. Cultural workers from different ethnocultural backgrounds recount stories of walks and bicycle rides through the parklands along the shore of Lake Ontario, the trails of Burrard Inlet, and the beaches of the Georgia Strait. All of these encounters with nature, whether in public space or private gardens, stimulate and catalyze "individual aesthetic creativity in the sense that creative workers exploit localised experiences as a source of ideas" (Drake, 2003, 515).

Proximity to nature and separation from city life are themes that characterize local artists' appreciation of suburban life, but they stand in marked contrast to the industrial, working-class heritage of inner suburbs like Etobicoke and North Vancouver. It is curious that cultural workers find inspiration in the natural features but not in the working industrial features of the local visual environment. In the central city, living and working in former industrial districts in reclaimed late nineteenth- and early twentieth-century factory warehouses with large floor-to-ceiling windows is an experience that is often glorified and romanticized by artists and upper-middle-class urbanites (Bain, 2003; 2006). However, in the inner suburbs, industrial landscapes, particularly those composed of mid-twentieth-century manufacturing and warehouse buildings with few windows, may be variously characterized by cultural workers as bleak, dry, flat, tired, and polluted. Inevitably, places "will be interpreted differently by different individuals and will provide different prompts and aesthetic raw materials" (Drake, 2003, 513). For the majority of cultural workers suburban industrial landscapes are far from being sources of "signs, ideas, and prompts" or places of romanticization (Drake, 2003, 517). There are only a few cultural workers in this research sample who seek inspiration in the poverty and decay of working-class suburban life.

Born and raised in the north end of Surrey, where he currently has a studio, one photographer has a strong emotional attachment to the area and the memories it shelters (Figures 4.4a and 4.4b). He has witnessed the landscape change from a residential European immigrant enclave of small working-class Hungarian, German, Dutch, English, and Italian homes around an elementary school and a community centre to an industrial zone of storage yards, auto body shops, warehouses, and factories with labourers from India, China, and Russia. This photographer has sought to document the socio-spatial transformation – the local poverty, the urban decay, and the encroachment of development on tiny pockets of residual space – and the land-use conflict

Figures 4.4a and 4.4b   The exterior and interior of a photography studio
in a converted industrial warehouse in Surrey (photographs by A. Bain).

this brings. His images have not been well received by the local arts establishment:

> They didn't like the fact that I showed the seedy part of the area ... I docu-
> mented whatever I saw, like junk. In some ways it's a very filthy area, but
> very beautiful too. They didn't want to see that side of it. They don't want to
> bring out any old memories of the decay down here. There's a lot of drug
> addicts down here living in old ramshackle houses. They don't want to see
> any of that ... They don't want any bad publicity for the area. They don't
> want a show that shows the decay or the people that are here. They want a
> politically correct show. (Photographer, Surrey, 24 May 2007)

While the curators of the Surrey Art Gallery may have liked this photographer's style, he maintains that they did not like the content of
his images. Unwilling to compromise his artistic vision, he has yet to
see these local landscapes of poverty and decay exhibited in Surrey.
His images offer insight into "the living city," accentuating the contrasts
between the everyday lived realities of people who have spent their lives
in Surrey and the imaging strategies of suburban cultural boosters who
often seek to accentuate the beauty of natural landscapes (Jonas & Ward,
2007, 170).

There are other cultural workers who rely on tangible and sonic
by-products of industrial suburbs as a valuable cover for their creative
labours. A North Vancouver stone sculptor (11 June 2007) explains
why he moved out of a shared studio in downtown Vancouver into an
architect-designed home studio and gallery where he has lived for the
last twenty-five years:

> I was looking for spaces, residential spaces within the suburbs of Vancouver,
> and this particular location was perfect because we back onto a heavy indu-
> stry area. I could build a studio which is in the back of the property and not
> in the neighbourhood because I'm not making any more noise than the
> industrial area behind us does. It was for this practical reason that we chose
> this location.

This sculptor could not find an affordable downtown live/work property
that would accommodate the mess, dust, and noise of stone sculpture.
He now lives in an inner-suburban residential neighbourhood just above
the main arterial road for industrial and commercial traffic that cuts
through North Vancouver's working port.

The interviews conducted for this research suggest that while many suburban cultural workers appreciate the natural elements in the physical landscapes that surround them, only a few value and explore the industrial lands and transportation corridors of working suburbia. The rusted and abandoned relics of industry may be a source of inspiration for some visual artists and a camouflage for a handful of others, but few cultural workers in this study sustain genuine interest in the lived realities of suburban industrial and manufacturing landscapes. Instead, suburban cultural workers express "urban ambivalence" steeped in "nostalgic pastoralism"; they idealize the suburb as the middle ground between the countryside and the city (Beauregard, 2006).

Suburban cultural workers, like the generations of middle-class residents who have gone before them, harbour "bucolic fantasies about moving to the countryside" while also maintaining contact with the city centre (Hayden, 2003, 22). Their views of suburbia have been distilled through a pastoral filter that frames the suburbs as a return to a simpler way of rural life that is thought to thrive in the countryside. This myth of urban pastoralism is deeply embedded in North American culture. The suburban project has long been driven by middle-class desires for easy access to unspoiled nature and open land. The rustic landscape, however, is inherently unstable and vulnerable to development pressures. In interviews, cultural workers treat the suburbs discursively as a rural substitute; they view them in a favourable light as places of space, greenery, tranquillity, and refuge. The desire to withdraw from society into an idealized rural landscape and the notion that a more meaningful life is possible with greater proximity to nature and distance from the constraints, complexities, and corruption of the inner city is a common theme in the interviews. Suburban cultural workers articulate a desire to combine isolation and community, nature and city, with a nostalgia for rusticity and an interest in urban diversity and complexity.

"Bourgeois Utopias"

Second to proximity to nature, in terms of reasons given by cultural workers for locating in the suburbs, is the quality of life that is perceived to be attainable. However, when data on housing are combined with quality-of-life data, this reason moves easily into numerical first place. The umbrella category "quality of life" in this book covers a range of characteristics of suburban living associated with middle-class domesticity: housing ownership, affordability, and size; convenience; security;

freedom; and privacy. For cultural workers, as for many middle-class North Americans from different ethnic backgrounds, suburbia remains an imagined landscape where "ambitions for upward mobility and economic security, ideals about freedom and private property, and longing for social harmony and spiritual uplift" are situated (Hayden, 2003, 3). In his book *Bourgeois Utopias*, Robert Fishman (1987, x), quoting Lewis Mumford, describes the suburbs as "a collective effort to live a private life" in a place deeply embedded with bourgeois culture and middle-class values. The earliest North American residential suburbs celebrated private property and the nuclear family. These bourgeois utopias of the affluent middle classes that idealized "countrified, quiet, and reserved suburban living" continue to influence how cultural workers in the present day describe their experiences of, and reasons for living in, the suburbs (Clapson, 2003, 7).[5]

Housing Ownership, Affordability, Size, and Convenience

In popular stereotypes, suburbs represent residential places where home ownership is high and population turnover low. Certainly, many of the earliest suburbs in North America developed in response to the strong homeownership aspirations of immigrant, working-class, and middle-class families. A home, as Richard Harris (1996, 89) documents within the context of the emergence of blue-collar suburbs on the periphery of Toronto in the early twentieth century, is "a physical presence; a setting for domestic activities; a site to which services may be delivered and from which jobs, shops, and open spaces are accessible. It is also a form of wealth." Like young couples who have for generations been pushed to the urban fringe where land values are cheaper and housing more affordable, many cultural workers see a move out to the suburbs as the most viable route to entering the owner-buyer housing market (Burnley et al., 1997). The housing experiences of an Iranian-Canadian mixed-media artist from North Vancouver (16 May 2007), as depicted in Figures 4.5a and 4.5b, illustrate this point:

> I moved to Canada in 1998 and my parents got here with my father's only brother in 2001. We invested in this house. It's a family income and savings and everything. We got it in this location because my Mum loved it and it was a really quiet, peaceful, family-type residence.

A desire to acquire property is not usually attributed to cultural workers, who are mythically expected to endure hardship in order to feed

Figures 4.5a and 4.5b   A family's property investment in North Vancouver with a studio constructed by a mixed-media artist in a backyard shed (photographs by A. Bain).

and embody their creativity. However, of the 57 cultural workers in the research sample, almost all of them owned a home, and 80 per cent of them had some kind of workspace in the home. Aside from a handful of rentals in apartment buildings and housing cooperatives, for the most part these homes took the form of owner-occupied bungalows, single-detached two-storey houses, and condominiums. Workspaces within these homes were variously crafted out of additional bedrooms, living rooms, basements, garden sheds, and garages, some created with the help of professionals and architectural plans and others merely with imagination and sweat equity.

A common interpretation of the suburbanization process is that suburbs are "the logical outcome of rational locators" reflecting the choice of households to seek "lower-cost accommodations and more efficient and less regulated landscapes, within a competitive urban environment" (Bourne, 1996, 179). This interpretation is reflected in the housing choices of the cultural workers interviewed for this study. A performance artist (16 June 2007) of 40 years who directed musicals at the Surrey Arts Centre in the 1970s long before he became a resident, explains why he moved to Surrey:

> I lived in Vancouver for years and we finally decided we were going to build a house and a friend of mine took me around. I had a figure in mind for what we were going to spend and I didn't want to leave the big city. I love the big city because all the action is here. We looked at houses but in the price range we were looking at, we could only get condos or a duplex. An old buddy of mine who is a salesman said we should look in Burnaby and New Westminister. We came across the bridge and suddenly I saved $30,000. I was looking for a house, front yard, and a backyard and I said, "Maybe I'd better rethink all of this." We ended up buying a house here in Surrey. Of course, I shifted my home and my loyalties in an instant.

In the suburbs, this actor could get more property and house for his money. He was motivated by housing price differentials and followed the consumer option of more house and garden space to be purchased. This actor-consumer is not alone in being lured out from the central city "to 'get on the property ladder' or to acquire a 'better property'" by the seeming greater range of homeownership possibilities available in the suburbs (Watt, 2009, 2880).

Bunting et al. (2004) have identified that the unaffordability of housing is a growing problem within the fastest-growing Canadian CMAs because these have inflated and exclusionary housing markets. High rental and purchasing costs within the central cities of Toronto and Vancouver have pushed some cultural workers out into the inner and outer suburbs where real estate is more affordable. As one long-time resident of south Etobicoke, a professional classical musician by training (24 October 2006), explains: "I like the idea that this house cost us $400,000 when we bought it, whereas if we purchased it in the Beaches, it would have been more like $750,000. It was reasonably priced out here." Attentive to comparable property values in the city, this musician, who plays with the Canadian Opera Company, is obviously pleased with the real estate investment decision he has made. As a homeowner with income sufficient to qualify for a mortgage, he is a consumer. The suburbs offer him and other cultural workers like him who earn (or who have a spouse or partner who earns) a reasonably regular wage "houses filled with comforts and conveniences, in communities of their own making and choosing" (Jurca, 2001, 8).

For those cultural workers who can see the imaginative possibilities in a built structure, and can provide the sweat equity and preliminary capital to transform it, the suburbs can provide a range of affordable housing options. As Figures 4.6a and 4.6b illustrate, a printmaker in south Etobicoke saw the structural potential in an abandoned building just off of the main commercial artery and transformed it into a live/work studio and gallery:

> The reason we bought in Etobicoke was that it was affordable. We were looking for a house at Jane and Bloor where we lived at the time in a flat and the house prices were just too expensive. We came to Etobicoke because the prices were less ... A friend of mine had driven by this building accidentally one day and she saw a for sale sign on it. She was an artist and we had talked about how we could get a better, bigger studio space for our future plans. She encouraged me to phone about this place, and it was affordable because it was derelict. It had been allowed to just mould. It had an oil tank that had been allowed to dry. All the pipes in the building froze and it flooded. It was basically a derelict building. Being derelict, it was affordable and I was crazy enough to buy it ... It was a lot of work to make it liveable. We did all of the demolition work ourselves and then I hired contractors. (printmaker, Etobicoke, 25 November 2006)

While not all of the cultural workers I interviewed were as risk-taking as this printmaker or as willing to contribute the hard, physical labour

Figures 4.6a and 4.6b   Sweat equity invested to convert an abandoned building in Etobicoke into a live/work printmaking studio (photograph 4.6a by A. Bain and 4.6b by L. Wheelans).

necessary to make an ambition of homeownership a reality, they were keenly aware of the financial savings to be had by forgoing a central-city location.

With increasing distance from the city centre, cultural workers were variously aware that the space they could have to live and to work in could increase and become more affordable. One well-established Canadian actress justified her residence in North Vancouver (4 June 2007) with the statement: "I happen to live in North Vancouver because the lots were bigger and the houses were more character-oriented and the taxes were lower." Another musician and storyteller explained that she moved from the Commercial Drive area in the east end of Vancouver into her current condominium in North Vancouver (21 June 2007) to get "a little more space in my home." How the additional space in a home is put to use obviously will vary by individual and by discipline, based on the kind of creative labour engaged in. Take, for example, the reason given by one Canadian composer of classical music for his need for additional space in his central Etobicoke home (9 November 2006):

> We have a large house. It's very long. We have studios on either end of the house so that we don't get in each other's way. You can't really do that downtown. I mean even the largest homes are large uprights. That's not any good because sound travels straight up and down, whereas long works very well. We can't really hear what each other is doing. Just the availability of that kind of structure to live in is very important.

Space in this instance affords a buffer of sonic solitude within which to create. Creative work often demands and induces total concentration. To be able to own, control, and organize the space of creative labour, to give it meaning with familiar objects and work rituals, is a luxury that many suburban cultural workers have been able to afford (Bain, 2007). Where significant portions of the creative economy are based on precariousness – low incomes,[6] unregulated work environments, and irregular employment opportunities – affordable homeownership opportunities can provide cultural workers with important experiences of control, stability, independence, and freedom.

A space of openness and freedom

The American architect Frank Lloyd Wright (1958) argues in his book *The Living City* that the biggest problem with cities is that they are too densely populated. He contends that the most appropriate settlement pattern to nurture freedom and creativity is one person per acre. Like Wright, Scott Thomas (1998) also maintains in *The United States of Suburban* that low population densities and spacious environments are a good fit with individualistic tendencies. Given the art world's celebration of individualism through bohemian mythologies, the suburbs would appear to offer cultural workers the space and the freedom to live where and how they please. In Australia, an appreciation of "the relative abundance of space in the suburbs" is emphasized by Gibson (2012, 252) following interviews with suburban cultural workers who associate spatial abundance with "freedom and an openness to experimentation."

However, the privilege of freedom does not extend to all social groups:

> To me the suburban environment is about space and about ease of access, although this area is becoming pretty unrecognizable in those terms. But that certainly is what it used to be about. We are still comparatively okay with safety concerns and all of the quality-of-life concerns. Hookers don't walk up and down my street. My impression of what the suburbs constitute is formed by my fifty years living here. I still remember seeing the first panhandler just here on Bloor Street across the road. You would just never see that up until about eight or nine years ago. There just weren't any up there. The suburbs for me are about space and access. (composer, Etobicoke, 9 November 2006)

Having grown up within, and returned as an adult to, an upper-middle-class neighbourhood in central Etobicoke, this cultural worker articulates conventional middle-class quality-of-life concerns about social changes in his residential neighbourhood. He perceives the suburbs as a safe haven, a place that should act as a buffer through the provision of social and physical distance from lower socio-economic classes and the urban ills of poverty, crime, and prostitution. But this perceived freedom to ignore social justice issues because they might distract from creative endeavours comes at a price, both for this individual and for society.

For other suburban cultural workers, it is a different kind of freedom that they associate with suburbia. It is the freedom attributed to life off of the cultural radar, in an untrendy part of the city variously referred to as "way out there" or "no-man's land" or "off the beaten track." In interviews, some cultural workers appreciate the suburbs for the sense of escape that the geographical and psychological periphery affords them, for the opportunity to get their hands dirty, and for the chance to make something from nothing. An Etobicoke photographer (8 November 2006) who relocated from New York City, and again from downtown Toronto, wants to capitalize on the untapped potential of the periphery:

> I still subscribe to that theory that if you build something unique, some people will come. They'll come locally and you could even draw some people from another place; maybe even downtown where everything is basically established. There's so many people that have already manifested these kind of ideas downtown, there's not much room for more. Whereas out here, this is sort of like the Wild West where you can start something.

This characterization of the suburbs as a cultural frontier emphasizes how suburbia can be a place where people value the pioneering qualities of individualism, independence, self-sufficiency, thrift, and risk-taking. Cultural edginess may be embraced for the sense of remove it affords from the institutional and social pressures "to conform to peer norms and expectations" in the competitive mainstream (Flew, 2012, 238).[7]

Privacy and Family Life

In North America, residential suburbs have been physically and socially structured to encourage privacy and separation. These are the very qualities that often directly appeal to the creative demands of cultural workers, who may alternate between social interaction and self-imposed isolation. One well-established visual artist, after completion of his post-secondary education, marriage to his wife, and the birth of his first child, bought a home in south Surrey that an architect renovated to meet his studio needs. He required:

> high walls, lots of light, north lights, skylights, and a sort of big open space and privacy. We were on 12 acres before moving here. I really like the privacy. When we came out here, we settled for one acre, which was the minimum. As you can see, we block the neighbours out with trees because you don't really want it to be disturbing. You don't want suburbia to totally rain down on you. (painter, Surrey, 3 July 2007)

A suburban location provides this painter with the space and privacy he needs in order to create and to raise a family. It also provides him with the opportunity to visually block out generic indicators of mass-produced suburbia and the presence of other suburban households. What has long enchanted middle-class residents about suburbia is not the social life but rather the exclusionary impulses it supports. It is quite common for people to know few of their neighbours and for suburban households to be separated by significant social distance (Baumgartner, 1988).

Many cultural workers choose to live and work in the suburbs as a compromise because it allows them to build families and to foster relationships within family units. "The suburb was not the best choice for an artist, but it was suitable to raise a family as a first step. This was my very first consideration" (visual artist, Mississauga, 1 November 2007). The suburbs are frequently framed as a safe place to raise a family:

> When I started thinking about having a family, with my husband, we thought "Where exactly do we want to be?" I had a real problem coming over the bridge. It felt so far away. We started looking around this area and started thinking we like mountains, we like hiking, and we like skiing. We go to Whistler and community centres on the North Shore. We wanted to raise children and to be central but have that feeling of a little bit of far away. It seemed that it perfectly fit with our lifestyle. I wasn't really thinking about my career in a way about how that fit into it. But we're so close to being central for my work and for my husband's work. We basically came and fell in love with the community of North Vancouver. We came here in 1994. It was a big move from downtown Vancouver. (actor, North Vancouver, 21 June 2007)

This actress portrays the suburb of North Vancouver as an ideal lifestyle choice for a young, active family who require access to the central city for employment yet desire suburban recreation and housing for their children. In negotiations with her husband, she made a comfortable compromise in moving to the inner suburb of North Vancouver. The suburbs allow many cultural workers to remain connected to family networks:

> I came to Mississauga because my entire family was based here. My parents were living there, my brothers and sisters and I moved here to Mississauga. I wouldn't live in any other town. Mississauga provides me with a link to where I come from ... I'm a second-generation immigrant. My parents came from India to Pakistan and from Pakistan we came to Canada. (visual artist, Mississauga, 25 October 2007)

"All my family is here" (theatre producer, Etobicoke, 19 November 2006) is a common refrain. For many cultural workers who give priority to family life and raising children, the suburbs offer a practical location to facilitate childrearing and caregiving within extended families.

Spatial Proximity

The residential and family orientation of suburban life does not preclude the use of suburbs for cultural labour. Almost half of the cultural workers interviewed regarded the in-between spatial location of the suburbs as conducive to the maintenance of central-city work connections. Cultural workers who live in the inner suburbs or who are more senior in their discipline and have developed extensive professional networks find it manageable to live in the suburbs but make a living elsewhere. As one classical musician (Etobicoke, 24 October 2006) explains: "You're close to the city. If you're a visual artist or a professional musician or an actor, it's easy for you to live out here and then get downtown for your gig, for your job." Further west, in Mississauga, cultural workers reiterate the ease of their commute to downtown Toronto: "I can get downtown to Toronto after 7 p.m. from my place (which is eight minutes from the Living Arts Centre) to Roy Thomson Hall in 20 minutes. It's nothing. People can get out here, but they have to drive" (conductor, Mississauga, 11 November 2006). That same commute would obviously not be so short during peak travel times.

Nevertheless, in Vancouver, residents of North Vancouver and Surrey also discuss the relative ease of access by car or by public transit to their bread-and-butter jobs or necessary cultural amenities and resources in the downtown core. A North Vancouver actress (21 June 2007) emphasizes how appropriate she finds the distance from her family life in the inner suburbs to her work in downtown Vancouver:

> I don't feel any dislocation. I feel very close. It is a really nice sanctuary to be away. This is my place. I don't have to be involved everyday with those people you know, artists out there. I have my place, my life, my family, all of that here to ground me. I really need that – my time, and myself, not as a theatre artist but as myself and I can't separate that. It's a really good distance. It would be too far away, if I'm on Bowen Island or Vancouver Island. In Surrey, I would probably feel like I'm not part of the community anymore, but this is just the right amount of distance.

Rather than fostering feelings of dislocation, spatial distance affords this actress a psychological buffer from professional demands and the "'headspace' ... to devote to creative work" (Flew, 2011, 5). Such distancing helps her to protect and nurture her creative self in the "sanctuary" that she calls the family home. Conversely, a well-established painter offers an alternative perspective on Surrey. He discusses the benefits of his home-studio location, accentuating how it is

> in very close proximity to the Vancouver airport and to Bellingham airport. It's very close to the border, so one can easily go across to receive or ship things or you have easy access to the U.S. It's a pretty good location. We're also close enough to Vancouver that when we need our urban rush, it's about 45 minutes away. Whether it's special movies or theatre or music or food or whatever, the urban life is there. (painter, Surrey, 3 July 2007)

For suburban cultural workers like this painter, journeys downtown for "an urban cultural fix" are expected. Such journeys are usually planned in order to maximize task completion and to minimize the time spent away from cultural labour:

> The North Shore is so close to Vancouver and I keep talking about going over the bridge to my work. I do go over the bridge quite often. You're so close to the downtown core and what's happening, by choice. I piggyback [activities] because of the traffic and the density. (potter, North Vancouver, 15 June 2007)

While a handful of cultural workers rely entirely upon public transportation to navigate the city-region, most suburban cultural workers interviewed either own, or have access to, a car or motorcycle so as to comfortably and conveniently negotiate the distances of the suburbs. The car is used by suburban cultural workers as a "management tool" to flexibly negotiate professional work routines and occupational expectations (Dowling, 2000). The car affords a sense of independence and freedom and helps to combat potential feelings of professional isolation. Car ownership increases travel mobility for cultural workers and facilitates access to, and use of, central-city cultural infrastructure.

Downtown Toronto and downtown Vancouver contain the majority of cultural facilities and organizations that suburban cultural workers in Etobicoke and Mississauga, North Vancouver and Surrey, rely on. Many visual artists, actors, musicians, dancers, and writers maintain memberships in downtown arts organizations so as to receive weekly newsletters and electronic updates on workshops, performances, and exhibitions.

Thus, both Toronto and Vancouver function as cultural information hubs for their respective city-regions. Cultural knowledge tends to be collected and made significant in the central city and then transferred to the suburbs through news media and the internet. This distribution pattern reinforces the perception that the central city is the creative focal point of the city-region.

The proximity of most suburbs to the downtown can be a double-edged sword. The central city provides valuable cultural sustenance, but it can also draw away cultural vitality from the inner and outer suburbs and foster cultural dependence rather than "urban-suburban interdependence" (Greenstein & Wiewel, 2000). The director of the North Vancouver Community Arts Council (29 May 2007) expresses frustration at the lack of local and regional awareness of the quality of suburban cultural products:

> There's lots of great product over here on the North shore, but people don't know it. They're still getting in their cars and thinking if they want to go to a good play, they've got to go downtown, when they don't have to. They don't have to go downtown to hear great music. They don't have to go downtown for great art.

A relationship characterized by dependency reinforces the misperception that if cultural activities occur outside the downtown core, they are less significant or are inferior.

The case study suburbs in this book began either as separate cities or have become cities. As one faculty member at Humber College in Etobicoke (4 November 2006) explains:

> Part of our mind knows that we can hop in the car and go downtown to visit the national this, the Canadian that, and have some amazing experiences and see some amazing work ... On the one hand, yes, we sometimes think that we're in the 'burbs and we have to go downtown for our cultural fodder. But by the same token because Etobicoke was its own city it had cultural activities happening here.

However, those same suburban cultural activities may get minimal recognition from central-city arts organizations, attract few audience members from downtown Toronto, and receive little local or city-wide media coverage.[8]

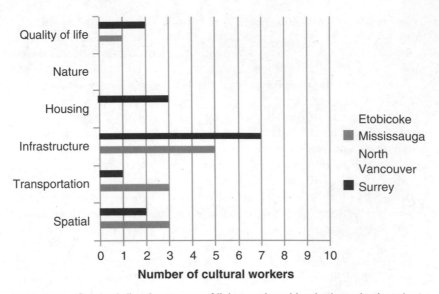

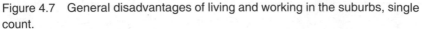

Figure 4.7   General disadvantages of living and working in the suburbs, single count.

Why Not Here?

Despite the positive spin on suburbia that cultural workers provide, several naysayers offer stories of the many challenges of suburban cultural production. Figures 4.7 and 4.8 summarize the general and particular disadvantages that cultural workers associate with suburban cultural practice. Although the major categories remain almost identical to those used for advantages (with the exception of employment replacing nature), the subcategories have changed to reflect unique obstacles identified by suburban cultural workers.

Cultural Infrastructure Deficiencies

The most commonly identified drawback to undertaking cultural labour in the suburbs is inadequate cultural infrastructure; this is the subject of more detailed discussion in chapters 5 and 6. Cultural infrastructure in its "hard" form refers to cultural facilities such as theatres, galleries,

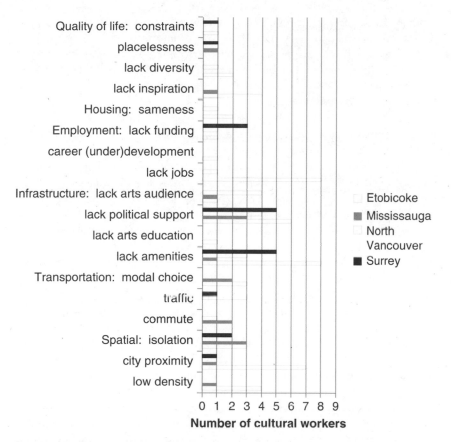

Figure 4.8   Disadvantages of living and working in the suburbs, multiple count.

studios, production centres, museums, archives, and arts education institutions. In its "soft" form, cultural infrastructure refers to arts audiences and information, knowledge, and communications networks that connect cultural workers and cultural organizations.

In interviews, arts administrators and professional cultural workers emphasized the hard cultural infrastructure shortfalls in their respective suburban municipalities. For example, the director of Arts Etobicoke (25 September 2006), the community arts council located in central Etobicoke, clearly states that there are very few venues in which to partake of cultural activity here. There are schools, there are churches, there are community centres, there are a few galleries, there are perhaps a few performing spaces and there are large multiplexes. But there are no museums. There are no large galleries. There are no performing arts centres. There is no concert hall. There is nowhere to go.

Her observations are echoed by a local photographer (8 November 2006) who has to go downtown for everything from foreign film screenings, exhibition openings, and concerts, to training workshops: "All the schools and colleges and universities are basically downtown … there aren't a lot of art galleries out here, private or pubic. There aren't a lot of unique music venues, like The Horseshoe." The shortage of spaces of cultural consumption is paralleled by a shortage of affordable rehearsal and performance spaces. As the director of Lakeshore Arts, the community arts council located in south Etobicoke (27 September 2006), explains: "There's really not a lot of space for local artists or local arts organizations to go and get something affordable that's an arts space. There isn't. It doesn't exist." Such infrastructural shortages apply not only to Etobicoke; similar comments are also made by suburban cultural workers about Mississauga, North Vancouver, and Surrey. A musician in Mississauga (18 October 2006) laments, for example, that the local arts council has a strong visual arts bias and that there are few "opportunities for musicians or for dancers … because there aren't any venues." A professional photographer in North Vancouver (22 May 2007) complains that "there's no support for photo-based artists here at all." In Surrey, an actor (16 June 2007) on the Board of Directors for the Arts Council of Surrey asserts:

Two things are lacking. One is rehearsal space … The other big problem culturally in Surrey, when it comes facilities, is gallery space. I don't think that's a unique problem to Surrey. I think most areas are short of gallery space always. There is never enough room for the artists to display their work or sell their work.

Such remarks may appear at face value to be discipline-specific, but they are a common refrain from cultural workers of all disciplines, and they speak to a fundamental dearth of suitable and financially accessible cultural venues in the suburbs.

How does this shortage of hard cultural infrastructure impact cultural workers? As will be discussed in subsequent chapters, it forces many cultural workers, like this visual artist in Etobicoke (4 October 2006), to adapt and to make creative use of space:

> I created a little studio in my house and started doing freelance work while I was being a mother and then I still always kept my hand in doing gallery work because that's what I really wanted to do. Because there weren't a lot of opportunities, I would show my work wherever I could: cafés, churches, libraries, anybody who would give me their wall to let me show my work. I would do art in the park.

In addition to the adaptive reuse of buildings and public spaces for producing and exhibiting art, there are other practical infrastructure considerations. A visual artist in Surrey (24 May 2007) recounts that "it's much more difficult for artists to find each other and find a market, or be able to do and successfully market their work. Everything is so much further and people in the city don't take you seriously because you're not living with them, you're not in there." In career terms, then, suburban cultural workers must often labour harder: to connect with arts audiences and their peers; to find outlets for their work; to obtain critical feedback, training, recognition, and financial support; to secure employment; and to acquire the resources necessary to undertake creative projects.

Although many suburban municipalities provide an adequate base of support for amateur and semi-professional cultural activities through educational institutions, non-profit cultural organizations, and governmental agencies, the real inadequacy lies in the lack of training for professional cultural practitioners. The remarks about the lack of cultural venues apply equally to professional development opportunities. An actress from Etobicoke (1 November 2006) complains that

> there is nothing in Etobicoke. There is no professional theatre company in Etobicoke. There is nobody out here that has a permanent professional home. There are people like me … There are dance studios and they are very good. There are lots of artists in the south Etobicoke Lakeshore area. If I wanted to train in dance, I could go to a local dance studio. But there is no professional theatre company out here. There is nobody out here running professional theatre classes, and part of the challenge is the same challenge

that I have found in running my classes. I may be working with many new beginning students, but my classes are professional classes. The challenge in running those kinds of classes is getting enough students to run each program if you don't have a lot of money for advertising.

Despite the presence of a handful of community colleges and university satellite campuses with arts programs across the case study sites, professional cultural workers frequently remark that they rely on the downtown arts scene for workshops, training sessions, job opportunities, and an arts audience.

At the beginning of chapter 3, I raised the challenge of connecting with suburban arts audiences from an arts administration perspective. Even with an array of financial and tactical resources for community outreach at their disposal, arts administrators often struggle to draw an audience, local or otherwise, to suburban arts events. A public outreach programmer (29 August 2007) for a public art gallery in Mississauga describes how

> there are these areas of mass population surrounding the gallery, but people just don't see it. They just can't find themselves in the gallery space. You have the Living Arts Centre at one side, the Square One shopping mall (that's really huge), and then this whole ring of condos around it. It's a bedroom community. Trying to get them out after a certain time and to get them to come into the gallery to participate in programming is something different.

In a suburban setting that can seem predominantly oriented to domestic, retail, and recreational pursuits, arts events can be perceived as disruptions of nine-to-five work and family routines rather than as pleasant distractions or intellectual stimulation. When suburban residents consume culture, it frequently involves commuting to the central city rather than partaking of local cultural fare: "There's a lot of North Shore audiences that still go to Vancouver for their cultural entertainment. We've got a huge amount to benefit from not competing with each other but really working collaboratively, and yet we're each trying to figure out what our pieces are" (arts administrator, North Vancouver, 15 June 2007). In North Vancouver, some arts administrators consider it important that local cultural venues individually determine their target audience and collectively work together so as not to cannibalize each other's audiences. Individual cultural practitioners may collaborate with colleagues to expand their audience base, but they rarely have the financial and temporal resources to extensively coordinate event schedules to maximize attendance. Suburban arts attendance is usually lower than that for similar events in the central city.

Suburban cultural workers and arts organizations usually have to work harder than their central-city counterparts to generate an audience. On the North Shore, for example, the audience for live theatre productions has been characterized by one actress as "a really tough crowd" (North Vancouver, 21 June 2007). She goes on to generalize that "we're not as culturally experienced or educated as a theatre-goer. It's not part of their life." In this quotation an "us-versus-them" distinction is made. On the one hand this actress considers herself a resident of North Vancouver, with the collective noun "we" (although she herself does not lack cultural experience or education), but in the next sentence her reference to "their life" suggests that the suburban life is different from hers.

Audience outreach is also affected by political and community support. Cultural workers and arts administrators in all four of the suburban case study areas seek political leadership in the cultural sector. In Etobicoke, a visual artist (5 November 2006) laments that

> there isn't a huge amount of cultural stuff happening. They don't close off streets to a street festival. It's hard to get the business community, the Business Improvement Associations, involved. They don't see the value of the cultural bottom line, which kind of reflects provincial and federal attitudes. "It has got to pay for itself" is what they think.

Many local businesses in Etobicoke, Mississauga, North Vancouver, and Surrey do, in fact, help to sponsor annual community arts events, but they rarely take the lead in proposing new cultural initiatives. New project proposals are usually left to cultural workers to develop, to underwrite with unpaid labour, and to lobby for financial support. The greater the place-branding potential of a proposed project, however, the greater the likelihood of business sponsorships and government support through grants. Even so, such arts support tends to wax and wane with changes in political administration. The director of the North Vancouver Community Arts Council (29 May 2007) bemoans such lack of support:

At the political level, municipally, it's abysmal right now. It really is. You'll get some people with vision in political power and then you lose them and you have to re-educate and rebuild. Or you hunker down and keep on doing the work until you get someone who comes along who has a creative vision.

Many arts administrators and cultural workers believe that the creative vision for a suburban municipality should be generated from the top down rather than from the grassroots up. Yet the visioning capabilities of politicians are frequently called into question; many are criticized for their economic bias and lack of appreciation for the complexities of cultural labour. A high-profile conductor in Mississauga (11 October 2006) insists that "there are lots of people politically who have not a clue of the value of arts in Mississauga." Another long-time Mississauga resident and sculptor (18 October 2006) directs his frustration more specifically at the mayor's office:

When you've got a mayor that doesn't care and a mayor that only cares about the bottom line, which is the dollar, it's really tough. We were at an event, the arts awards, the other night, and she came and talked, and over and over again she said, "It's up to you people to promote it." I just came from a place in Argentina that's a fairly poor country, a city of 250,000, and the city and the government all sponsor and support this and it really all comes from the top. If they show the example that we want to have artwork, then the people will see it.

This sculptor wants a mayor who leads by example, a mayor who actively places art and artists high on the funding and policy agenda for her or his term in office, and does so with an eye to long-term community development potential as opposed to short-term economic gain. His desire for long-term planning and sustained investment in the cultural sector by local politicians echoes sentiments expressed by suburban cultural workers across the country from Ontario to British Columbia.

Isolation, Distance, and Cultural Elitism

Infrastructural inadequacies combined with a paucity of transit connections in the suburbs can contribute to a sense of professional isolation among suburban cultural workers.[9] Following interviews with suburban cultural workers in Melbourne and Brisbane, Gibson (2012, 253–4) notes how "suburban creative practitioners often suffer from isolation and disconnection, failing to recognize their potential either personally or in the wider social and economic contributions they might make." An 18-year-old former recipient of the Mississauga Emerging Visual Arts Award states:

In my opinion, each arts organization works as an individual unit, lacking promotional synergy, collaboration, and community engagement in the arts. There are many creative minds willing to make a difference, working on new proposals and ideas, but still the suburban context is artistically isolating and lacks the assistance, critical mass, and opportunities for fulfilling the potential of the arts community and myself individually. (visual artist, Mississauga, 1 November 2007)

Isolation increases the logistical challenges and the entrepreneurial skills needed to undertake cultural work, in terms of self-generating access to suppliers, resources, markets, venues, audiences, and employment opportunities (Bennett, 2010). Networking with colleagues is central to career development in any profession, but when undertaken from a suburban location, it can cost cultural workers more in terms of time, money, and effort. Writing about the experiences of creative industries workers in the outer suburbs of Australian cities, Felton et al. (2010, 61) emphasize how distance was perceived as a "perceptual and geographical barrier." Distance "made attending formal professional networking activities less likely and appealing; it also meant that clients and perhaps more importantly *potential* clients from the inner city were less likely to conduct business with one located in the outer suburb" (Felton et al., 2010, 62). Similar networking and client-recruitment obstacles can be found among Canadian suburban cultural workers. In the suburbs, there are often fewer physical spaces and opportunities to meet with other professional cultural workers, thus limiting the opportunities for face-to-face contact and the possibilities for extending one's influence beyond local, place-based ties and relationships.

While on tour following the publication of her second book, Canadian poet Brenda Schmidt, who lives in the small mining town of Creighton, Saskatchewan, was repeatedly asked whether she felt isolated living and working in such a small community, so physically removed from the heart of the literary scene. In an article published on the Canadian literary website *Bookninja*, Schmidt (2005), tongue in cheek, reflects on why "writers [are] expected to flow to larger and larger centres to oxygenate their creativity and pump up their opportunities." She confidently asserts that a small-town location affords her the physical and mental space, time, and silence within which to work and to see ideas through to completion with minimal administrative interference. Appreciation for the space, time, and quiet afforded by a more remote location is also expressed by suburban cultural workers. While some individuals embrace suburbia because it distances them from the pressures and competition of the mainstream and opens up the creative possibilities that such a spatial buffer affords, others are less optimistic about the impact of remoteness.

The perception of cultural and physical distance between suburbs and the central city, as well as within suburbs themselves, is distorted and varied. Local cultural workers who own cars usually regard the central city as a close commute and as a valuable source of cultural resources. But even with access to private transportation, the distance across a given suburb can seem vast. That sense of suburban spatial expansiveness and inaccessibility is only heightened by inadequate public transit:

> Everything is so far away and the public transit is crap. If you don't drive, it is really hard to partake in cultural experiences because it's a hassle. For me, it's actually easier for me to get to Toronto and commute throughout Toronto or walk throughout Toronto, whatever it is, than it is for me to get to somewhere at the other end of Mississauga. (visual artist, Mississauga, 30 October 2007)

For this artist, the cultural opportunities of the central city are perceived as closer and more accessible than their local suburban counterparts. The converse rarely holds true. Few central-city residents perceive the inner and outer suburbs as close and accessible. Instead, they are frequently described as "an out-of-the-way location" (actor, Etobicoke, 1 November 2006). In the experience of one Etobicoke printmaker (25 October 2006), "you talk to someone about where you have your studio, they'll say 'Way out there?' and I'll say 'The Queen streetcar goes right by the end of my street.' There is that perception of it being sort of far away."

In conversation, the suburban case study sites are frequently casually referred to as "out there" on the urban periphery, in a cultural wasteland. "People in the city see the west end, see Etobicoke, as no man's land. If you say to someone at Yonge and Bloor, 'Let's go to Lakeshore and Kipling' they'll say 'Are you insane? Why would I go there? Is there anything there?'"(director, Lakeshore Arts, 27 September 2006). In the public imagination suburban locations can be regarded as cultural voids.

Working on the geographical margins, on the edge of the city, can foster feelings of insignificance, silencing, and disrespect relative to central-city cultural production. The manager of the Assembly Hall performance facility in Etobicoke (13 November 2006) explains that "when you're talking about the challenges of being near a large cultural centre, there is certainly a feeling of, if it's in Etobicoke, it's probably not that good." In many instances, suburban cultural workers recount experiences of having their work "openly patronized as second-class artistic practice by the professional arts community" (Ford-Smith, 2001, 21). Descriptors such as: "insular," "peaceful, cozy and very family-friendly," and making "very nice looking, healthy [local] cultural news" (sculptor, North Vancouver, 16 May 2007) are frequently used to characterize suburban cultural production. For example, the curator of a photography gallery in North Vancouver (20 June 2007) emphasizes that the cultural community of peers that he deals with on a daily basis is in Vancouver and that there are few local suburban photographers working at the "highest level" whom he would profile with an exhibition:

> Often people find it off the beaten track to come over here and see exhibitions. You're perceived to be on the periphery, which is not so much geographical as psychological. You are just lacking constituency really when you're outside an urban centre. As much as there is a kind of support for our existence in North Vancouver, there's not a lot of enthusiasm among people who might be culturally sensitive because a lot of people just simply don't understand what it is we're doing. I don't mean that to be patronizing. The constituency of people who appreciate what it is we're doing are, by and large, based in Vancouver and outside in other cities. It's just the nature of it.

In this quotation, the assumption that cultural knowledge is held by those who live and work in the central city and that "real" art is practised and appreciated in downtown studios and exhibition and performance spaces conveys some of the arrogance and élitism that suburban cultural workers encounter.

A newly hired curator for a gallery in Mississauga casually admits in his interview that "there are two artists in Mississauga who are producing work that is actually relatively interesting" (curator, Mississauga, 30 August 2007) – but only two! This statement implies that the bulk of work produced by local artists is anything but interesting, and he describes it as "technically and conceptually naive." On the receiving end of similar snap judgments about quality and expertise, one visual artist describes her experiences volunteering on a committee with the Toronto Arts Council:

> I found there was a very bad attitude towards people in Etobicoke. People in Toronto thought that people in Etobicoke were just bumpkins and that we didn't know how to do anything ... I have found that Toronto is kind of snotty that way. They think they know how to do everything and the rest of us need to be taught. (mixed media artist, Etobicoke, 4 October 2006)

In the face of such attitudes that appear to be blatantly dismissive of the work and skill sets of suburban practitioners, it can be an ongoing challenge to confidently exude a sense of self as professional cultural worker. Cultural elitism is very clearly at play in the landscapes, institutions, and lives of suburban cultural practitioners. Misjudgments and misperceptions about the quality of creative work and the technical abilities of creative labourers add an additional layer of complexity to the challenges of suburban cultural production.

Conclusion: An Emergent Relationship

The relationship between cultural workers and suburbs is organic and emergent. This is a give-and-take relationship that depends on what an artist chooses to accept and to reject. For most of the cultural workers in this study the suburbs are "less an object to be contemplated than a medium that they inhabit" (Perl, 2001, 23). The suburbs are less a source of inspiration than a lived reality. With suburban habitation come both opportunities and challenges. On balance, the advantages are more numerous than the disadvantages, demonstrating that the majority of suburban cultural workers value the suburbs positively as a place to live and work.

This chapter has shown that rational and emotional locators influence the decisions of most suburban cultural workers to live and work in the suburbs. Like the traditional middle class, suburban cultural workers prioritize the need to be close to nature and the countryside. Green spaces, waterscapes, and farmers' fields provide the illusion of rural simplicity, solitude, and escape from the city, in the knowledge that urban entertainment, education, and employment amenities are available on demand, with a commute. The suburbs are framed as "bourgeois utopias" as much for cultural workers as for the middle class. Cultural workers emphasize the same quality-of-life preferences as the middle class: affordable and spacious homes in safe, family-oriented neighbourhoods that provide domestic convenience, freedom, and privacy. In contrast to David Ley's (1996a) portrait of artists as quintessentially urban people who lead the new middle class in their residential choices, this chapter has shown that cultural workers are embedded in the social and physical fabric of suburbs and that they value suburban life for the same functional advantages as do the traditional middle class. With their often quite conventional residential location choices, suburban cultural workers are just as likely to be followers as leaders of the middle class.

The attractiveness of the suburbs to creative people has been contrasted with the everyday challenges. This chapter has identified the primary drawback to suburban cultural production as the inadequate quantity, quality, and diversity of cultural infrastructure and the isolation and stigma that derive from such deficiencies. The following chapters in this book examine these infrastructural constraints in greater detail and consider how suburban cultural workers have creatively and proactively responded to them.

Chapter Five

Suburban Cultural Infrastructure Planning Formulas: From District to Multiplex

This chapter explores some of the gaps and complicating factors in the provision and planning of cultural infrastructure in Canadian suburbs. It begins with a preliminary consideration of what cultural infrastructure is and of the role of the public sector in investing in and inventorying cultural assets. Attention then turns to the expansion of the municipal cultural planning movement in Canada and the part it has played in strengthening the voice of culture at the infrastructure-funding and policymaking table. Particular consideration is given to how cultural infrastructure is addressed in the cultural plans of Toronto, Mississauga, Surrey, and North Vancouver. An analysis of these planning documents reveals an almost formulaic drive by suburban municipalities to overhaul the management of the cultural sector and to replicate the cultural-district model of central cities. Two case studies, of the Mississauga Living Arts Centre and the Surrey Arts Centre, illustrate how this suburban quest to achieve the agglomeration and the place-branding benefits of proximity finds spatial expression in the construction of flagship cultural multiplexes.

This chapter argues that suburban municipalities too readily gravitate towards celebrating culture in its consumptive and spectacular form and attend inadequately to the practical infrastructural challenges of supporting local cultural production in diverse and dispersed contexts. This chapter also asserts that suburban municipalities should strive to be complementary and locally focused in providing cultural infrastructure; they should shift attention away from showcase, city-wide facilities to the decentralized delivery of cultural spaces for professional, emerging, and amateur cultural practitioners at the neighbourhood and district spatial scales.

Cultural Infrastructure

Public infrastructure is usually understood by municipal decision makers as the core engineering assets of a community (e.g., bridges, roads, sidewalks, sewers, water supply, waste disposal, public transit, and parks). These assets are expected to operate smoothly and efficiently, and, wherever possible, to remain silent and invisible in the landscape. Clean drinking water, an efficient road network, a reliable waste disposal service, and an affordable public transportation system are all considered essential components of a sustainable community where residents can expect to experience a good quality of life. A modern urban planning framework is concerned with the prevention or correction of conditions that could hurt people or property. Health and safety, accessibility, and shelter are considered important qualities to be protected in the management of public infrastructure. A rational approach to planning, firmly rooted in pragmatic functionality, is unlikely to register culture and cultural resources as essential public infrastructure. Take, for example, the infrastructural learning curve that the manager of arts services for Surrey (18 June 2007) charts for her colleague, the city manager:

> If you can imagine working with a city manager for a number of years whose background was engineering (whose cultural background was not Canadian) and when he first started coming out to the opening of recreation centres, he even wondered why we were building those. I guess his notion was that they should be private and they should be provided for by the private sector. It was a huge learning curve for him. He really came around in terms of understanding the value of leisure services to the community and how that fit in with roads and sewers and hard infrastructure that he was more familiar with. But it still was taking a long time for him to grasp the importance of the cultural aspect – of how we can contribute, not just to the quality of life but also become part of that whole economic generator within our city. Whether it's through cultural tourism or just creating a great place to live and therefore attracting businesses and more residents to the area, to creating a more positive sense of place and dealing with social issues.

That culture and infrastructure are two words that can go together and productively inform each other is a recent realization in some suburban municipalities. In some urban planning and policy documents cultural centres and libraries are classified as community recreational infrastructure and thus remain bureaucratically hidden as cultural resources. But "infrastructure should not exist just for the sake of expediency, it should become the pretext for art and social welfare" (Ingersol, 2006, 20). Cultural infrastructure, as the manager of arts services for Surrey thoughtfully articulates, is particularly well positioned to simultaneously address the potential artistic as well as the social welfare functions of infrastructure.

To generate a national dialogue and a collective understanding of what cultural infrastructure is, Nancy Duxbury (2008) and her research team at Simon Fraser University's Centre of Expertise on Culture and Communities (CECC) have undertaken a valuable preliminary study that summarizes the current state of knowledge about Canada's built cultural environment. Their report, *Under Construction: The State of Cultural Infrastructure in Canada,* synthesizes the findings of five regional roundtable discussions, an international forum, and a survey of cultural administration and planning practitioners. The authors begin by broadly defining cultural infrastructure as "physical assets and spaces whether they are full-time or part-time, single purpose or multi-purpose, historic or contemporary, that have cultural products and accommodate and satisfy the requirements of cultural activities and cultural industry" (Duxbury, 2008, 17). Such a wide spectrum of cultural infrastructure includes visible and less-visible components for public interaction (e.g., performance, interpretation, exhibition, debate) and facility support (e.g., creation, rehearsal, production, storage, live/work, preservation). This definition of cultural infrastructure recognizes that the spaces used for cultural activities in any given municipality are many and varied. Such cultural spaces can extend well beyond traditional defined-cultural-use facilities to include multifunctioning features of the urban built landscape as well as multi-use outdoor, recreational, heritage, and transportation venues.

Throughout the report, the authors emphasize that reliable and comparable data on cultural infrastructure in Canada are limited and that inventories produced through cultural facilities mapping exercises are fragmented and inconsistent. Of particular relevance to the suburban context, Duxbury identifies substantial knowledge gaps and invisibilities around small, emerging, and unconventional spaces. Such spaces are invaluable to small and mid-sized arts organizations, yet they are often unrecorded and unrecognized in formal records. There is no publicly available, nationally comparable inventory of Canadian cultural facilities, nor is there a comprehensive inventory assessment process in place.

The greatest increase in cultural infrastructure inventory occurred in Canada in the decade following the establishment of the Canada Council for the Arts in 1957. To commemorate Canada's Centennial year, the federal government financed the construction of community centres and single-purpose cultural facilities, such as theatres and concert halls, across the country. The Centennial investments in culture made the 1960s the decade with the greatest growth in the number of cultural facilities (Duxbury, 2008). But many of the Centennial buildings have reached the end of their life cycle. An aging building stock brings higher operational and maintenance costs as well as issues of safety, comfort, and acoustics. Without sufficient capital investment in improvements to extend the lives of these buildings, the value of these cultural assets has decreased since the 1990s and will continue to do so.

Over the last two decades, much investment in cultural infrastructure has come from the private sector. "Starting in 1991, the private sector provided 93 per cent of all capital investment in the Arts, Entertainment, and Recreation industry; while this share declined to 53 per cent by 2006, the private sector remains a significant player" (Duxbury, 2008, 49). Other key players include the federal departments of Canadian Heritage, through the Cultural Spaces Canada program initiated in 2001, and Infrastructure Canada. With a one-third, two-third respective split, these two departments have together contributed approximately $543 million to cultural infrastructure projects between 2001 and 2008 (Duxbury, 2008).

Before the establishment of the Cultural Spaces Canada program, the federal government inflicted deep cuts to cultural spending. A moratorium on capital infrastructure expenditure by the Department of Canadian Heritage resulted in a significant gap in the physical construction of cultural infrastructure between 1993 and 2001, particularly in smaller communities outside large cities (Duxbury, 2008). The Cultural Spaces Canada program, renamed the Canada Cultural Spaces Fund in 2009 with funding extended until 2015, helped to address this gap. With a mandate to "improve physical conditions for artistic creativity and innovation," the program "supports the improvement, renovation, and construction of arts and heritage facilities and the new acquisition of specialized equipment as well as conducting feasibility studies" (http://www.pch.gc.ca/eng/1267728945673#a1). The Cultural Spaces program has been in operation a little over a decade. However, it can take that long to bring a cultural infrastructure project to fruition. Thus it is important to have stable, long-term funding horizons that can support the infrastructure planning process through election cycles, so as not to endanger the public legacy of cultural facilities.

Another significant hurdle in the development of cultural infrastructure in Canada is the collaboration of all three levels of government. Current cultural policy and programming frameworks have been characterized as uncoordinated, fragmented, and not maintained for a sufficient duration to permit long-term planning (Duxbury, 2008). New construction project priority-setting for cultural infrastructure investment often occurs at the upper levels of federal and provincial governments, while the long-term costs of sustaining operations for the life cycle of a structure usually fall to municipalities. The devolution to municipalities of responsibility for the operational costs and maintenance fees of cultural spaces has left many cities unable to afford to renew or replace an aging and deteriorating stock of facilities.[1] Municipalities must contend with competing and elemental infrastructure investment demands. Stories of crumbling bridges, collapsing highway overpasses, neglected parking garages, ruptured water mains, and proliferating potholes are a steady source of journalistic fodder. Sometimes deteriorating public infrastructure inspires works of art (Chung, 2009), as in the photographs of Montreal artists Claudia Ficca and Davide Luciano that playfully reimagine potholes as pedicure footbath, fishing hole, pet cemetery, baker's cauldron, baptism vessel, beer cooler, and swimming pool (www .mypotholes.com). But more often than not, infrastructure deficiencies are less a source of creativity or amusement than of frustration for residents and politicians alike as municipalities struggle to set funding priorities to meet immediate needs and balance investment in different types of infrastructure.

In suburban municipalities where laissez-faire government policies have allowed new low-rise, car-dependent office, commercial, and residential development to continue at a fast pace, transportation and recreation infrastructure demands have superseded those of culture for many decades. On the City of Mississauga website, for example, at the top of the alphabetical list of "important locations throughout the city that may be of interest to residents" are baseball and softball locations (www.mississauga.ca/portal/services/facilities). In this suburb, a map of baseball diamonds is prioritized over maps for hospitals, fire stations, and the courthouse. On the online interactive flashmap of Mississauga facilities, the number of "cultural sites" (9), major community centres (12), community halls (8), and libraries (18) is numerically far surpassed by recreational sites. The number of fitness centres (7), arenas

(14), rinks (3), pools (18), splash pads (18), tennis courts (48), golf courses (2), leash-free zones (7), and picnic parks (10) totals 127, easily exceeding the officially designated sites of cultural infrastructure (www .mississauga.ca/file/COM/facilitymap_ecity_2007_08_27a.swf). In addition to such official sites of cultural infrastructure as theatres, museums, galleries, and arts centres, the indoor and outdoor recreational spaces in Mississauga are used for arts and culture programming (e.g., Lakefront Promenade, Civic Square, Library Square, Queen Elizabeth II Gardens, and city-owned arenas and community centres).

There are eight theatres in Mississauga with a total of approximately 2,500 seats; this is one-third the number of theatre seats found in either Edmonton or Winnipeg (City of Mississauga Culture Division, 2009). Although the number of theatres and theatre seats are useful infrastructure statistics, many of the local community theatres are poorly signed and have little structural presence in the urban landscape. For example, "Meadowvale Theatre is a large and very good facility, but from the outside it just looks like a utilitarian building. It's not even a signed property. It doesn't have horticultural displays. It doesn't have the signage that it truly deserves" (curator, Mississauga, 5 October 2007). While there is clear signage for parks and recreational facilities, many cultural facilities and attractions in Mississauga have not been made equally as visible to the general public.

There are four community-owned and city-owned museums in Mississauga. In terms of population size, other comparable Canadian cities (e.g., Edmonton, Vancouver, and Winnipeg) have four to 10 times as many museums (City of Mississauga Culture Division, 2009). There are six public art galleries and 16 commercial art galleries in Mississauga; both Vancouver and Winnipeg have twice as many public art galleries, while Edmonton has double the number of commercial galleries (City of Mississauga Culture Division, 2009). Aside from the Art Gallery of Mississauga or the Blackwood Gallery at the University of Toronto Mississauga campus, most galleries in Mississauga tend to showcase traditional paintings and sculptures, leaving local visual artists who practise different art forms to look elsewhere for inspiration. As one curator (Mississauga, 30 November 2007) explains: "If I want to see good art that's really interesting to me, I'm more likely to find it in one of the hundreds of galleries in Toronto rather than one of the 20 galleries in Mississauga. I think a lot of people feel the same way, where it's easier or even cooler to go to Toronto." In Mississauga, there are only two post-secondary institutions – the University of Toronto Mississauga campus and the new Sheridan College Hazel McCallion campus (located behind the Living Arts Centre) – to augment cultural training and dialogue (Figure 5.1), whereas other cities of a similar size might have at least six. In addition to a shortage of galleries and post-secondary educational institutions, Mississauga has only two communal centres for the practice of visual arts (e.g., Visual Arts Mississauga and the Living Arts Centre); Edmonton has 12 such centres and Vancouver and Winnipeg each have 25. These statistics suggest that culture remains secondary to recreation in Mississauga.

In the City of Surrey, British Columbia, culture is gradually achieving greater recognition from city councillors as an essential valued service,[2] but the provision of hard infrastructure remains to be seen as a funding priority. In a municipality that emphasizes fitness and active living, where "parks embody the image and identity of the city" (PERC, 2008, 55), funding has been directed towards the construction of community recreation centres (7), swimming pools (12), ice arenas (5), athletic fields (224), golf courses, and motorsport racetracks. Sports facilities and sporting activities in Surrey have traditionally had a higher status and have attracted more attention at the funding table than cultural facilities and events (Dunphy, 2009).

In Surrey, the culture portfolio is the purview of the Department of Parks, Recreation and Culture, a department that has a general mandate to enhance the quality of life of the community. One component of this mandate directly references culture, with a goal to "preserve, develop, and deliver cultural, informational, and educational resources and services" (http://www.surrey.ca/3464.aspx). Other dimensions of the mandate are concerned with celebrating diversity and community identity, ensuring accessibility and inclusivity, and promoting individual and community wellness – all of which could apply to culture as well as to recreation. Certainly, a recent revamping of the city website is helping to direct greater attention to cultural facilities and services in Surrey. A dedicated culture and recreation web page now profiles the "arts in Surrey" with the tag line: "Surrey is home to an active arts community, performing arts venues, a contemporary art gallery, an annual children's festival, a growing public art collection and more" (http://www.surrey.ca/culture-recreation/1517.aspx). Such attention to the arts is a recent phenomenon.

In his book *Indy Dreams and Urban Nightmares* Mark Lowes (2002) recounts how, in 1997, when Molson Indy Vancouver (MIV) organizers were looking to relocate the motorsport event from its downtown site at

Figure 5.1   Sheridan College's new Hazel McCallion campus in Mississsauga (photograph by L. Wheelans).

False Creek, Surrey Mayor Doug McCallum (1996–2004) wrote to Indy officials inviting the event to his municipality. In his bid to host the event, McCallum argued that Surrey was an ideal site because it had a large land base, wide streets, strong community support, and a volunteer base experienced at hosting the Greater Vancouver Open (GVO). Quoting Mayor McCallum, Lowes (2002, 71) writes: "The GVO puts Surrey on the map in golfing, because it gets broadcast all over North America and also gets satellite coverage to other countries – it puts Surrey on the map internationally. Golfing is a major aspect of tourism today; it's a huge draw if your community has a well-known course, and if you have a PGA event to go with that, it only boosts your image." Major sporting events like the GVO and the MIV were seen by Surrey political and business élites as effective image-promotional vehicles whose benefits outweighed the necessary investment of public money in infrastructure costs. A public statement made by the mayor emphasizes the commitment of growth boosters to hosting sports entertainment megaprojects: "The citizens of Surrey have been extremely supportive of our bid for the Indy. I think we delivered the message loud and clear that *Surrey is open for business*, and city council and staff are ready to do what they can to create local jobs and economic growth" (Lowes, 2002, 73).

Although ultimately unsuccessful in their bid to lure the Molson Indy to Surrey, the attempt demonstrates that city officials are willing to work hard to accommodate the revenue needs of the sports industry. "Meanwhile, support for sectors that [could be] much bigger generators of sustainable economic activity [are] neglected because they have a lower profile or the competition is less visible" (Turok, 2005, 33). A political willingness to lobby for culture, and to support it with the same enthusiasm and financial investment as sport, is gathering momentum in Surrey.[3] The voice of culture at the infrastructure funding and policy-making table in municipalities is building with the growth of the cultural planning movement in Canada.

Municipal Cultural Planning

Cultural planning can be defined as an integrated approach to urban development that encourages municipal policy makers to strategically use tangible and intangible cultural resources to leverage social and economic development opportunities for a community (Bianchini, 1993). In Ontario, the Ministry of Tourism and Culture has been a strong advocate of cultural planning. In 2003, a working group was established to organize 10 provincially funded Municipal Cultural Planning forums between 2005 and 2007, headlined by consultants Greg Baeker and Glen Murray. The forums targeted municipal staff, elected officials, and local cultural, business, and community leaders to encourage a broader vision of culture and cultural resources as pillars in new economic development and community-building strategies. At these forums, municipal cultural planning was presented through case studies of good planning practice as an activity that is horizontal, inclusive, integrated, and ongoing. The objectives of the forum were: to present cultural planning as an activity that helps to break down silos separating different forms of local cultural activity; to build and strengthen local and regional networks; to share experiences and good practices; and to identify needs and opportunities to advance cultural planning in Ontario.

A significant by-product of the Municipal Cultural Planning forums was the formation of the Municipal Cultural Planning Partnership (MCPP). As detailed on its website, MCPP is a coalition of provincial government agencies, municipalities, cultural service organizations, and post-secondary institutions that "have come together out of a shared belief in the power of culture to transform local economies and communities" (www.ontariomcp.ca). MCPP presents itself as the primary promoter of cultural planning in Ontario.

The Creative City Network of Canada (CCNC) plays a key role in facilitating the sharing of cultural planning information among professionals working in municipal cultural service delivery. Founded in 1997 with support from Vancouver's Office of Cultural Affairs, the CCNC established itself as a not-for-profit organization in 2002. Information is disseminated through its website (www.creativecity.ca), a listserv, and an e-newsletter (*Creative City News*) as well as at annual conferences and workshops. The CCNC has also published two instructional toolkits, *Cultural Planning Toolkit* (Russo and Butler, 2006) and *Cultural Mapping Toolkit* (Stewart, 2006). These have been widely disseminated and supported by workshops such as the one below, electronically advertised at the Martin Prosperity Institute at the University of Toronto in March 2010:

> Discuss. Learn. Connect.
>
> Join cultural planning practitioners from across Ontario for an informative workshop on cultural planning and mapping. Presentations on Creative City Network of Canada's cultural planning and mapping toolkits, panel discussions from those in the field and a facilitated discussion on what's next will give you the knowledge and connections to move the cultural planning agenda forward in your community.

The toolkits, workshops, and networking opportunities have served to streamline and standardize the cultural planning process, producing a formula that is relatively easy for municipal Parks and Recreation departments, Community Services departments, and Urban Planning departments to follow.

Many of the suburban cultural planners and arts administrators interviewed for this study lauded the CCNC as a valuable source of useful and current cultural policy information and best-practice examples. In the words of the Cultural Development Coordinator for North Vancouver (9 June 2007), who regularly attends the annual conference of the CCNC:

> The vision for the Creative City Network is incredible, in keeping the dialogue going across the nation and being able to transfer best practices and the learning around the country without having to reinvent the wheel. It is an amazing resource, both for municipal staff and for practitioners and service agencies ... The most useful thing for me is the website resource which has been refreshed and updated regularly. It's my first point of research

when I'm looking for something and I usually find the tools that I need there for subsequent searches ... I'm usually way too busy and I'm always impressed and amazed by how generous the community is. They provide a newsletter that keeps everyone up to date and they have an annual publication. The research that they do and then publish is impressive. The most useful thing that they've done recently is to provide toolkits for cultural planning and cultural mapping.

Local suburban cultural planners turn to the CCNC for cultural policy that is easily appropriated. Their searches for policy information are often unsystematic and rarely involve a "sophisticated process of critical comparative analysis and contextual evaluation, adaptation, and contestation" (Stevenson et al., 2010b, 167). The mobility and translatability of cultural policy are minimally questioned by local cultural policy practitioners. In a decade, the CCNC has positioned itself as one of the primary providers of cultural policy information in Canada; it has become an online cultural administrative hub.

In Canada, cultural planning for an entire municipality is usually taken on under the remit of a cultural policy adviser and a steering committee, who, in turn, often outsource to consultants. Two key dimensions of cultural planning are cultural mapping and community participation in the development of a cultural plan. Cultural mapping, internationally recognized by UNESCO as a tool for preserving cultural assets, involves the systematic identification and inventorying of local arts, culture, and heritage resources (Evans & Foord, 2008). Once cultural data about a place have been collected, categorized, and organized in a database, they can be displayed in maps using geographic information systems (GIS) software. Such maps render cultural assets more easily marketable by a municipality, with the maps themselves taking on agent-like roles in urban planning debates (Stevenson et al., 2010a). This form of mapping has become one of the most common "benchmark activities" used by Canadian municipalities to geographically represent cultural infrastructure and lay the groundwork for developing a cultural plan (City of Mississauga Culture Division, 2009).[4]

A cultural plan works within a medium-term planning window of five to 10 years. It may take more than a year to complete and over a decade to initiate. Through community consultation, a municipality's arts and cultural resources are assessed to identify what makes each place distinctive and what is needed to augment this cultural distinctiveness. Like most plans developed by professional urban planners, a cultural plan is a general guideline for planning action that operates as a vision of what could be and establishes strategies to realize that vision. In the words of cultural planning consultants Franco Bianchini and Lia Ghilardi (2007, 284), a cultural plan in its ideal form provides a framework around which local "governments and stakeholders alike can build coordinated actions, investments, policies, and communications." A cultural plan becomes a public statement of a community's recognition of the arts as an essential service, a statement that details future cultural development priorities.

The earliest and most extensive commitments to cultural planning in Canada were made by two cities that had the financial and human resources to direct towards refining the management of their cultural portfolio, Vancouver and Toronto. In the 1980s and 1990s, municipal decision makers in these cities began to develop policies, tools, and plans to assume responsibility for cultural development. Vancouver and Toronto, with their respective incarnations of Cultural Services departments, became models of cultural action to which other municipal cultural service providers looked for ideas for shaping their own programs and policies.[5] As suburban municipalities on the outskirts of these large CMAs have matured, they, too, have sought to better manage their cultural responsibilities through cultural planning practices.

The degree of commitment to a cultural plan and the finances to realize its vision vary among municipalities. Inevitably, there is a danger that a cultural plan is just another trendy policy document that has little leverage and few teeth, a document that is generic rather than unique, created out of expectations that the cultural planning process should be undertaken because every other municipality is doing it. While imitation might be the highest form of flattery, the cynic might suggest that cultural plans have become just one more task to complete in the creative capital formula of urban-economic development.

The suburban municipalities examined in this study are at different stages in the cultural planning process, their progress charted online through posted reports. This ready availability of reports suggests a desire to publicly demonstrate commitment to the development of cultural policy over time. Commissioning more studies and reports that chart progress in the refinement of the delivery of municipal cultural services seems easy and affordable relative to the larger task of investing in new cultural infrastructure, hard and soft. Certainly, there is a need in the municipal management of culture for visioning, community dialogue, and public accountability, but after a decade of doing this one would expect that these practices would have begun to deliver more tangible benefits to cultural practitioners and arts organizations. It would appear that, whether or not a municipality has a cultural plan, there is minimal impact on the lived reality of cultural production within its environs. The challenges of cultural production for suburban practitioners remain constant. As the following examples of cultural planning initiatives in Etobicoke, Mississauga, North Vancouver, and Surrey demonstrate, the greatest shift in the contemporary cultural landscape of municipalities has occurred in the amount of resources directed towards the public administration of culture.

Etobicoke

There is no cultural plan specifically for Etobicoke. To appreciate the impact of cultural planning on this inner suburb, the wider metropolitan context must be examined. Toronto's first attempt at a cultural plan was released in 1994 under the title *Metro Culture Plan: Redefining Our Cultural Framework*. This report included the arts, heritage, libraries, and cultural industries in a life-cycle view of culture that recognized the linkages between the creation, production, distribution, and conservation of cultural products. The result of a two-year public consultation and discussion on the cultural responsibilities of Metro Toronto, this report revealed challenges to the future of cultural development. The first challenge identified was

> a lack of cultural facilities and programs outside the core of Metro. Most of Metro's cultural infrastructure is located within the central area of the City of Toronto. While this concentration of cultural resources has helped to create a vibrant downtown and should be maintained, a lack of cultural facilities and organizations outside of the core limits participation by many of Metro's residents. (Municipality of Metropolitan Toronto, 1994, 11)

The report went on to cite the success of the North York Performing Arts Centre (renamed, post-amalgamation, the Toronto Centre for the Arts) as an indication of "audience support for facilities outside of the central area" (Municipality of Metropolitan Toronto, 1994, 43). In anticipation of future Metro-wide investment plans to rectify cultural facility imbalances, many metropolitan municipalities completed feasibility studies for cultural facilities. It is within this context that brief mention is made of the soon-to-be inner suburbs of Etobicoke, Scarborough, and North York. The City of Etobicoke built the Nielsen Park Creative Centre in 1992 and hoped to augment its two other small arts exhibition facilities, the Etobicoke Civic Centre Art Gallery and the Franklin Carmichael Art Centre, with a new multi-purpose community arts centre. Such a multi-purpose facility, the Assembly Hall, was eventually opened by the City of Toronto in 2001.

Although it strives to develop an integrated cultural policy across Metropolitan Toronto, Toronto's first cultural plan makes no mention of the cultural policy landscape of Etobicoke or of any of the other municipalities within the Metro partnership. Yet culture was precisely one of those areas of service provision requiring cooperation and coordination between the two levels of government because both Metro and local municipalities had cultural policies and programs in place and there was no legislative framework to guide the division of services.

Two years after amalgamation, in 2000, the newly appointed City of Toronto Council requested a draft cultural plan from the likewise newly created Culture Division (a subdivision of the Economic Development, Culture, and Tourism Department) to guide the city's cultural development in the coming decade. *The Creative City: A Workprint* (City of Toronto Culture Division, 2001) was drafted as a discussion document and provided the foundation for the *Culture Plan for The Creative City*, which was adopted by city council in 2003. City council was ambitious with their request; they wanted a cultural plan that would "position Toronto as an international cultural capital and enhance the role of culture in the economy and the social life of the city" (City of Toronto Culture Division, 2003, 49). The final document seems less of a plan and more of a glossily imaged and flamboyantly narrated sales pitch for Toronto's "cultural renaissance" and its potential as a creative city. The 10-year plan emphasizes talent and tourist attraction through attention to cultural facilities expansion, facade improvements, heritage preservation, community festivals, and public art.

As part of the cultural planning exercise, the Culture Division commissioned a cultural facilities review. This review culminated in the production of a Cultural Facilities Database of 750 cultural operations in the city, including galleries, museums, libraries, community centres, assembly and concert halls, school auditoriums, and churches that are either publicly or privately owned (City of Toronto Culture Division, 2003). The database, in conjunction with an accompanying report, *A Map of Toronto's Cultural Facilities: A Cultural Facilities Analysis* (ERA Architects Inc. et al., 2003), helped reveal the extent of Toronto's cultural facilities. The database was divided into four categories: hubs, incubators, showcases, and cultural memory sites. Hubs referred to non-traditional performance and exhibition venues that provide community-driven cultural activities; there were 532 of these, 60 per cent of them in the downtown core. Incubators referred to artist-run facilities that function as "a hothouse for the creation of new ideas, approaches and productions" (ERA Architects Inc. et al., 2003, 7); there were 292 of these, the majority of them clustered in specific "urban neighbourhoods." Showcases referred to market-driven facilities with a regional, national, or international profile that are leaders in cultural education and research and that function as key tourist attractions and destinations; there were 122 of these. Culture memory sites referred to facilities that support culture as a heritage resource and function as repositories of collective memory; there were 238 of these. When facilities played more than one role they were identified as multivalent complexes. The report divided the city into four districts (North, East, South, and West), corresponding to the municipal boundaries before amalgamation.

The report showed that 67 per cent of cultural facilities (503) are located in the South District, which is similar to the boundaries of the former City of Toronto, often in high-pedestrian traffic areas near public transportation and major road access (ERA Architects Inc. et al., 2003). Of the remaining three districts, the North, East, and West contained 90, 90, and 72 cultural facilities respectively. Of the 72 cultural facilities in the West District, which corresponds to the boundaries of Etobicoke, 58 were hubs, 21 were memory sites, 16 were incubators, and there were no showcase venues. Clearly, there is a strong cultural facilities bias in favour of the downtown core. Even when added together, the former municipalities of North York, Scarborough, and Etobicoke total only 252, half the number of the central city.

The report also revealed that three-quarters of Toronto's cultural facilities (predominantly incubators and showcases) were privately owned. The 31 facilities that Toronto "owns directly or has invested in – by way of direct grants, tax exemptions, below-market rent ... or development agreements – are generally in poor physical condition" (City of Toronto Culture Division, 2003, 18). The built fabric of cultural infrastructure in Toronto and its suburbs, as in other municipalities across the country, is deteriorating.[6] In the estimation of the report's authors, "the health of the City's cultural infrastructure and the delivery of the City's cultural objectives rests strongly on partnerships with the non-City-owned facility sector" (ERA Architects Inc. et al., 2003, 5).

In the report, a cultural facilities map generated using 2002 data was overlaid with a demographic map to determine where resources were most needed and where "arts districts" were emerging. The mapping exercise showed that cultural facilities are located "in clusters and corridors and are not evenly distributed across the city" (ERA Architects Inc. et al., 2003, 2). In the downtown core, the map showed cultural corridors of foot traffic between cultural institutions with the potential to be branded as arts districts. Yet no mention was made in the cultural plan of what the mapping tool revealed about cultural asset shortages on Toronto's periphery. The photographs and the text in Toronto's cultural plan celebrate the city's culture as a downtown phenomenon. Ironically, despite amalgamation and the documentation in 1994 and 2003 of an inequitable distribution of cultural facilities and arts programs across the region, the shortage of suburban cultural infrastructure remains largely unaddressed more than 16 years later. Two follow-up progress reports to the cultural plan, one released in 2005, the other in 2008, perpetuate this legacy of suburban erasure. These reports are little more than laudatory statistical summaries of cultural plan indicators and supposed achievements that use a repetitive cultural portfolio of central-city events, organizations, and facilities as representative of the health of the cultural sector across the city-region. A more recent snapshot of the cultural resources available in Etobicoke and used by the cultural workers interviewed for this study is provided in Figure 5.2.

Mississauga

A clearer picture of the impact of cultural planning on the suburban periphery can be achieved by examining suburbs that are municipalities in their own right. In the City of Mississauga, decision makers in 1991 expressed a fleeting interest in developing a cultural plan. It has taken nearly two decades to realize this ambition. In 2005, the Mississauga Arts Review Taskforce was convened by city council to determine the needs of arts and culture organizations. The task force recommended that the city develop an "arts strategy" that would inform future municipal planning. The city hired the consulting group Canadian Urban Institute, who worked in conjunction with cultural planning consultants Authenticity, Artscape, and Novita to run stakeholder discussions and workshops with local arts and culture groups, community organizations, businesses, and residents, from which a cultural plan was generated. In June 2009, city council approved Mississauga's first *Culture Master Plan*. The cultural plan is presented as a "short-term vision focus[ed] on building a cultural infrastructure that is community-driven and neighbourhood focused" (City of Mississauga Culture Division, 2009, 5). The ambitious intent

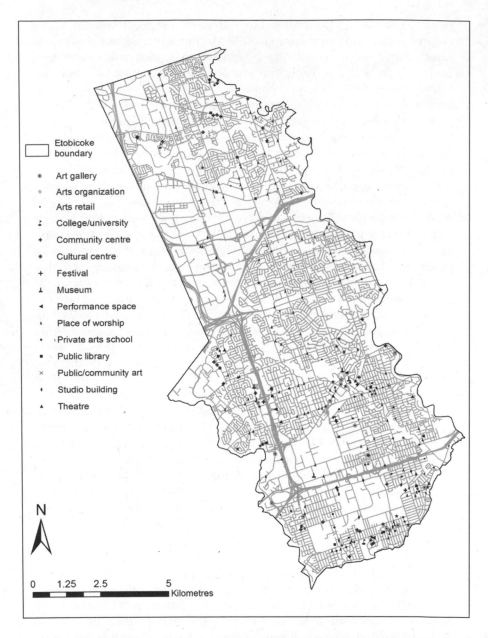

Figure 5.2   Cultural resources in Etobicoke, 2010 (map by W. Sturm).

of the document is to outline how "Mississauga can transform into a culturally significant Canadian city" by strengthening arts, culture, and heritage organizations and providing greater opportunities for citizen participation in cultural activities (www.mississauga.ca/portal/discover/ artsandculture).

During the public consultation phase, residents repeatedly emphasized the need for easier access to local cultural information. In response, the city has undertaken a cultural mapping exercise that documents art, culture, and heritage resources, activities, venues, organizations, and practitioners, and displays them on an interactive online map. Designed to run on all platforms and mobile devices, Mississauga's Culture on the Map went online in July 2012 (www.cultureonthemap.ca). A virtual "hub for all things cultural," the map allows users to search by keyword, category, or neighbourhood to locate and obtain contact information (email, websites, and social media) for cultural assets (Robinson, 2012). The cultural planning formula was reversed in Mississauga in that the cultural mapping process was a successor to, rather than a precursor of, the cultural plan.

The cultural planning exercise highlighted a number of critical findings with respect to cultural infrastructure. First, "there appears to be no rationale for cultural facility distribution overall within the city" (City of Mississauga Culture Division, 2009, 8). Second, "there appears to be no rationale for how cultural resources might be deployed in individual neighbourhoods either as amenities or as contributors to a critical mass of complementary uses which created a desirable destination" (City of Mississauga Culture Division, 2009, 8). Third, cultural planning is integrated in a piecemeal fashion into public administration, in that the Culture Division, Recreation and Parks, Planning and Facilities, and Property Management departments all have mandates that deal with some aspect of cultural infrastructure and cultural resources. Fourth, "cultural facilities in Mississauga have no communications platform for interface with the public and no forum for dialogue among themselves" (City of Mississauga Culture Division, 2009, 8). In sum, Mississauga is "underserved by cultural infrastructure and the existing cultural infrastructure is poorly distributed relative to where people live and the communities with which they identify" (City of Mississauga Culture Division, 2009, 62). Phrases such as "no rationale," "not well-integrated," and "general lack of awareness" all convey a clear message that the cultural sector in Mississauga has not been adequately planned or managed since the city's inception. The intent of the five-year cultural planning vision for Mississauga is to establish a framework for addressing some of these identified cultural infrastructure shortcomings.

The consultants for Mississauga's *Culture Master Plan* maintain that, given Mississauga's edge-city role in the Greater Toronto Area, it should allow Toronto to host the flagship cultural institutions for the city-region and instead concentrate on providing locally focused cultural opportunities for its residents. They persuasively assert: "This does not prevent regional cities from having their own flagship facilities such as the Living Arts Centre ... Indeed it does not preclude cities like Mississauga from developing signature cultural attractions or major tourist destinations along waterfronts or emerging urban nodes. But it does mean that local engagement and participation in culture should be the first priority and that no large-scale projects should weaken the city's commitment to community based cultural infrastructure" (City of Mississauga Culture Division, 2009, 34). Investing in cultural resources in the unspectacular, small-scale, and local over the spectacular, large-scale, and regional is a valuable and significant recommendation, but it is one that may not appeal to city councillors intent on branding and marketing their municipality. City councillors may worry that such a strategy may cause their municipality to be seen regionally and nationally as second-tier rather than first-rate, the understudy rather than the main cultural act. Mississauga, like many other suburban municipalities, has chosen to focus on the management and communications side of the cultural sector equation.

In addition to the community dialogue initiated by the cultural planning process and the placement of culture on the political radar, the main outcome of the cultural plan has been to bolster the public administrative and management side of cultural production. The Office of Arts and Culture in Mississauga has been renamed the Culture Division (to suggest a broader cultural remit) with the mandate to give grants and to assist local cultural workers with their promotion and marketing efforts. Within this new office, a manager has been hired to look at policy development and to consider how recommendations from the *Culture Master Plan* can be implemented. For now, increasing public awareness, strengthening the flow of information about culture, and "creating a network of cultural partnerships" remain high on this municipality's cultural planning agenda (City of Mississauga Culture Division, 2009). In addition, the city's website provides an online annual Culture Planning Calendar that lists major community events and provides links to some of the larger community organization websites. Figure 5.3 illustrates

Figure 5.3   Cultural resources in Mississauga, 2010 (map by W. Sturm).

the range of cultural resources that cultural workers identify and use in Mississauga.

Surrey

In July 2008, the City of Surrey adopted the *Parks, Recreation, and Culture Ten-Year Strategic Plan* developed during the previous year by the consultancy firm Professional Environmental Recreational Consultants Ltd (PERC) as an update to the 1999 *Cultural Strategic Plan*, which was grounded in the 1996 *Parks, Recreation, and Culture Master Plan*. The 1996 plan provided the framework for much of the investment in culture that has occurred in Surrey over the last decade, including an expansion to the Surrey Arts Centre, the construction of the Bell Performing Arts theatre, and the construction of the Surrey Museum and Archives. The only recommendation not implemented is the decentralization of arts and heritage services into community recreation complexes. The 1999 plan focused on increasing the promotion and marketing of culture in Surrey. It led to the creation of a *Cultural Marketing Plan* in 2005 as well as to the establishment of a Youth Arts Council and a Public Art Advisory Committee.

The 2008 cultural plan development process involved reviewing reports and an inventory of parks, open space, facilities, and programs; collating results from a survey questionnaire completed by nearly 1,200 residents; undertaking community stakeholder interviews and focus groups; and hosting six community forums and two public open houses. The plan makes 57 recommendations (of which 25 cannot be funded within existing per capita spending levels) under five general "strategic directions," two of which specifically address existing and new infrastructure demands. The bulk of the infrastructure discussion attends to recreational needs with a remarkable degree of specificity. For example, there are recommendations to construct three covered sports courts and additional fitness facilities, invest in outdoor pools, build synthetic turf surfaces, replace outdated playgrounds, expand city-wide trail systems, address dog-walking and regulatory enforcement issues, create new community horticultural displays, implement an integrated pest management plan, and revise the parkland acquisition strategy. Buried within this eclectic infrastructure discussion are eight recommendations that address cultural service concerns (e.g., enhancement of the public art program and improvement to arts and heritage initiatives).

In the cultural plan, the chapter on cultural services opens with the assertion that Surrey has invested heavily over the last decade in arts and heritage services. Yet it then goes on to state that "there is a feeling among many arts groups, in particular, that a great deal more is needed to rebalance public attention and investment, in order to recognize the importance of these service areas, and bring them up to a level of prominence and support that has been enjoyed by the city's 'physical recreation' services" (PERC, 2008, 72). An imbalance between recreational and cultural infrastructure is a widely felt and long-standing frustration among cultural workers and arts administrators. The primary cultural infrastructure need identified in the cultural plan is for Surrey to "provide arts and crafts studios, heritage display spaces, and performing arts rehearsal spaces" (PERC, 2008, 71). Consultation with the Surrey Arts Council revealed the lack of accessible community arts spaces, particularly artist-run spaces, in each of the town centres. It also emphasized the usefulness of a cultural and arts inventory to determine gaps in service provision. The consultants responded with the suggestion that gallery, rehearsal, and studio space could be added to existing community centres incrementally in the coming years as the opportunities presented themselves. One example of a recent response to these recommendations is the addition of the Surrey Urban Screen to the west wall of the Chuck Bailey Recreation Centre. This projection site is an outreach venue for the Surrey Art Gallery, visible from the Sky Train between the Gateway and Surrey Central stations and the City of Surrey now advertises it as "Canada's largest permanent, outdoor urban screen" (www.surrey.ca/culture-recreation/7315.aspx).

In a discussion of cultural infrastructure trends, the cultural plan productively identifies the challenge of "centralized facilities trying to deliver decentralized services in a city as physically large as Surrey" (PERC, 2008, 74). The report's authors go on to suggest that "addressing decentralized and distributed services in arts and heritage will require creative partnerships and new delivery systems in the future, including expanded partnerships with local arts organizations. Opportunities for decentralized cultural facilities (e.g., exhibition space) may also be achieved through the development process" (PERC, 2008, 74). This theme of decentralization is not unique to Surrey; it is an elemental cultural infrastructure challenge faced by suburban municipalities across the country. A common neoliberal response to this challenge is to emphasize the strategic management role that public-private partnerships can play in overcoming physical distance and hard cultural infrastructure shortages. Figure 5.4 provides an indication of the cultural resources available to cultural workers in Surrey and their spatial distribution.

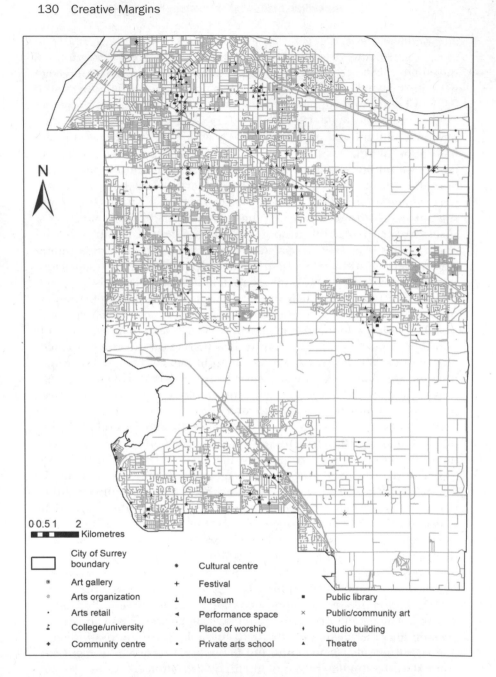

0 0.5 1   2
■ ■ ■ ■ ■ ■ Kilometres

◻ City of Surrey
  boundary   ⊛ Cultural centre

⊛ Art gallery   + Festival

⊙ Arts organization   ⊥ Museum   ■ Public library

· Arts retail   ◄ Performance space   × Public/community art

⌁ College/university   ▲ Place of worship   ◆ Studio building

✦ Community centre   • Private arts school ▲ Theatre

Figure 5.4   Cultural resources in Surrey, 2010 (map by W. Sturm).

North Vancouver

On the western side of Metro Vancouver, the District and City of North Vancouver have together developed a cultural plan. The District and City of North Vancouver are distinct entities, each with its own mayor and council and administrative structure. There are, however, several areas of administrative overlap where the municipalities consider it efficient and appropriate to combine service delivery (e.g., police and fire services). Cultural services, as recommended by the first cultural plan, undertaken in 1989, are delivered collaboratively. Much cultural planning in North Vancouver over the last two decades has focused on refining the delivery model of arts funding and program support structures.[7] A new arts administrative structure was implemented in 2007 and the North Vancouver Office of Cultural Affairs (NVOC) was established to manage all arts funding and cultural development services for the District and City of North Vancouver. In 2010, NVOC changed its name to The Arts Office and launched a new online cultural events calendar *on!* and website (www.artsoffice.ca).

The cultural plan currently in use in the District and City of North Vancouver was prepared by a steering committee and accepted by councils in 2002. The cultural plan was developed in two phases. The first phase laid out goals and strategies, while the second phase focused on the development of community-level partnerships to help implement the "strategic directions" envisioned by the cultural plan. Seven goals were identified: to strengthen alliances and partnerships for effective management and growth of the arts and cultural sectors; to expand opportunities for young people to experience the arts; to increase the variety and diversity of arts and cultural experiences; to provide spaces for cultural and artistic activities; to improve communication and information resources for the arts community; to expand the market for arts and culture; and to further develop and sustain a thriving cultural economy. North Vancouver has since established project teams that focus on the development of opportunities for youth, a coordinated approach to marketing, and a centralized information centre that supports the production of a biannual report to council. No concrete plans have been made to invest in new facilities. The intent is to maximize the usage of existing spaces and those in development (e.g., Lynn Valley Library and Town Centre) by making them more affordable to arts groups, by fostering organizational partnerships, and by inventorying arts resources (venues, skills, and grants). Interestingly, in its cultural planning documents,

North Vancouver explicitly recognizes the value of fostering spatial flexibility, particularly the use of non-traditional venues for cultural purposes (e.g., libraries, churches, and parks), and has sought to realize the
potential of the internet as a cultural promotion and networking tool.

Unlike Surrey, Burnaby, or Richmond, North Vancouver does not run
a municipal gallery facility with significant hands-on curatorial direction.
Instead, the bulk of its cultural funding, approximately $200,000, is put
into a community arts assistance grant program. Since 2000, attention
has also been directed towards establishing a public art program and
policy; there are now over 80 pieces of public art in the collection of the
city and district. Figure 5.5 documents the diversity of cultural resources
in North Vancouver.

While focused on their own cultural programming and policy initiatives, North Vancouver cultural planners have consistently remained
aware of, and are actively involved in, cultural planning initiatives at
the wider, regional scale. In 1994, an inter-municipal committee of
staff with cultural planning responsibilities approached the Greater
Vancouver Regional District (GVRD) for endorsement of, and support
for, a strategic cultural planning process for Greater Vancouver municipalities. In 1995 the GVRD authorized the process and the Regional
Cultural Plan Steering Committee was formed. A regional cultural
strategy was eventually drafted, but, without buy-in from all participat-
ing communities, it was never implemented. Instead of disbanding,
group members have continued to meet because "there have been so
many benefits from meeting and for exchanging their practices and
ideas and keeping in touch with each other" (cultural planner, North
Vancouver, 17 June 2007). Many municipalities remained active participants in the regional group, meeting approximately eight times a
year in Burnaby until, in June 2009, a new Regional Culture Committee
(RCC) was formed. The RCC has membership from North Vancouver,
Port Coquitlam, Richmond, Surrey, and Vancouver. A review of committee meeting minutes reveals that it has given out over $100,000 worth
of cultural grants for region-serving arts and culture organizations
and focused on developing a regional approach to culture promotion. Leading up to the 2010 Winter Olympics, attention was directed
towards how best to access some of the $20 million in Olympic legacies
funding.[8]

The regional cultural strategy in development by RCC emphasizes a
Web-based delivery program centred on a regional cultural calendar.[9]
The RCC's (2009, RCU-17 and RCU-18) goal is to "create an integrated

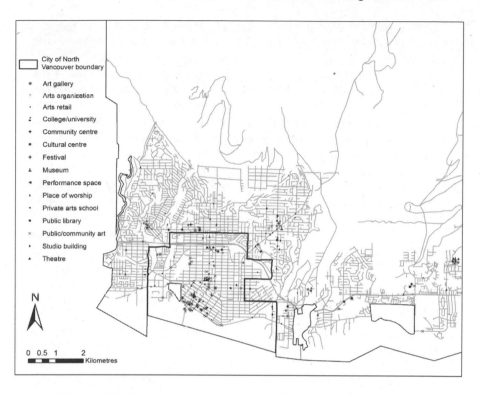

Figure 5.5   Cultural resources in North Vancouver, 2010 (map by W. Sturm).

regional cultural infrastructure" by building "a centralized arts and culture hub that provides easy access to a wide range of activities and events,
both professional and community-based, comprising a calendar of events,
workshops, and networking opportunities." This new project highlights
the growing recognition in suburban municipalities of the importance
of investing in the virtual and technological dimensions of cultural infrastructure to better unite communities and to help overcome some of the
challenges of physical distance.

Urban Creative Milieux, Cultural Quarters, Cultural Districts, and Cultural Nodes

Across the country from the heartland to the hinterland, from large and
small urban centres to rural areas, from central cities to suburbs, the
distribution of cultural infrastructure is uneven. A hierarchy of access to

cultural spaces exists. Cultural planners, through cultural mapping and cultural facilities inventorying, seek to document, although rarely rectify, this imbalance.

The largest concentration of prestige cultural facilities, cultural districts, cultural organizations, and cultural practitioners is usually clustered in the downtown cores of Canada's largest cities.[10] Policy makers and scholars of the cultural economy frequently assert that such clustering of the cultural and creative industries occurs in places that possess an "urban edge: that is, a mix of old and new buildings, an active streetscape, mixed use, contemporary design, cafés and bars, nightclubs and … 'the evening economy'" (Montgomery, 2008, xxv). Such heritage, mixed-use developments, places of encounter, and street life that extends into the evening are assumed to be absent in the suburbs.

Thus, it is cities and the centres of cities that retain the greatest density of cultural infrastructure. It is cities that remain privileged as the singular, inspiring locations of cultural innovation, progress, and creativity (Nicolaides, 2006; Perl, 2001), and celebrated as the places where "risks are taken, problems are raised, experiments tested, ideas generated" (Urban Cultures Ltd, 1994 cited in Montgomery, 2008, 31). The importance of urban culture as a primary force in shaping cityscapes is evidenced by the terminology that has developed in the fields of urban planning and the cultural economy to celebrate culture's significance and concentration in particular places: "creative milieux," "cultural quarters," and "cultural districts." This terminology underscores the widespread belief among scholars of the cultural economy that creativity is grounded in particular geographic, social, and economic contexts (Mizzau & Montanari, 2008). As Kevin Stolarick and Richard Florida (2006, 1802) write, "the general creative milieu of a place with a prominent presence of artists, musicians, and other creative people increases overall creativity and innovation by providing stimulus and inspiration." They go on to assert that "the roles of density and diversity in all of this are to compact all of these people into a space where they collide and, hence, where … important interactions and spillovers can occur." Urban cultural policy in post-industrial cities has come to be dominated by this Floridian idealization of the value of creative clusters as sites of capital accumulation, cultural innovation, and social inclusion (Oakley, 2004; Van Heur, 2009; Gospodini, 2009; Curran, 2010; Flew, 2010; Heebels & van Aalst, 2010; Martí-Costa & Pradel i Miquel, 2011).[11]

Drawing on the work of Swedish urban historians Gunnar Törnquist (1983) and Ake Andersson (1985),[12] the British creative-city consultant Charles Landry (2000, 133) writes that a creative milieu "is a physical setting where a critical mass of entrepreneurs, intellectuals, social activists, artists, power-brokers, or students operate in an open-ended, cosmopolitan context and where face-to-face interaction creates new ideas, artifacts, products, services and institutions, and as a consequence contributes to economic success." The wealth-generating potential of creativity, as intensified through urban places that possess certain tangible and intangible qualities, has garnered significant attention from academics and decision makers alike in the post-Floridian urban marketplace.

In an effort to kick-start urban revitalization initiatives and as "a panacea to implement broader city expansion and regeneration plans" (Evans, 2009, 1003), Western city leaders frequently seek to identify and brand neighbourhoods with a concentrated presence of cultural activity within their jurisdictions as "cultural quarters" and "cultural districts" (e.g., gallery district, theatre district, film studio district, and design district). "Cultural quarter" is the term that is popular in the United Kingdom, while "cultural district" is the term most commonly used within North America (Brooks & Kushner, 2001). In urban planning discourse, both cultural quarters and cultural districts are creations of government that speak to a particular understanding of cultural production and provide official ways to visually order and curate the city (Keith, 2005). Such branding generally requires a mixed-use area of the city with a variety of cultural venues of different sizes where space is set aside for artistic production and consumption and where cultural facilities and programs function as anchors to attract and sustain the cultural economy. Highlighting many of these attributes, Simon Roodhouse (2006; 2009, 82–3) presents a useful definition of cultural quarter as "a geographic area of a large town or city that acts as a focus for cultural and artistic activities through the presence of a group of buildings devoted to housing a range of such activities and purpose-designed or adapted spaces to create a sense of identity, providing an environment to facilitate and encourage the provision of cultural and artistic services and activities." To use the language of cultural economists, cultural quarters or cultural districts are defined by "the non-random concentration or agglomeration in space of differently sized autonomous cultural firms or institutions, with a prevalence of small and medium ones" (Mizzau and Montanari, 2008, 652). More than just a local artistic scene, the key attribute of cultural quarters in these definitions is the urban spatial clustering of cultural producers for economic benefit.

Roodhouse (2009) maintains that cultural quarters usually develop in a place where a cultural presence already exists in embryonic form. Mizzau & Montanari (2008, 667) insist that exploiting this "fertile, endogenous creative 'humus'" should be an essential public policy directive (although this is often not the case). Montgomery (2008) outlines a typology for developing successful, planned cultural quarters. There is little room in his discussion for a consideration of the significance to their success of the organic and unplanned qualities cultural quarters. Assertive top-down planning may limit rather than liberate, privileging economic, marketing, and hard infrastructure priorities over bottom-up community-building and informal interaction (Power and Scott, 2004).

From an urban planning perspective, Montgomery (2008, 348) asserts that cultural quarters in Western cities should contain strong representation by local cultural industries.[13] Sectors in the cultural industries include the performing arts, the visual and craft arts, the literary arts, music, television and radio broadcasting, film, video, photography, fashion, industrial design, and architecture. Each sector has its own production chain and economic profile, and is usually dominated by major organizations at the national and international levels. The sectors themselves are often volatile and in constant flux as creative workers come and go, and regulations, competition, consumer preferences, and patterns of demand change.[14]

Once the presence of cultural industries in cultural quarters is assured, the "aim should be mixture and diversity, nooks and crannies, concentrating in particular on small-scale venues, workspaces and complementary activities" (Montgomery, 2008, 348). Montgomery details how set programs can help to initiate the process by focusing on: the development of a range of cultural venues; the provision of space for artistic production; the use of creative industries as business incubators to foster economic development; the animation of the public realm with public art, festivals, and events; the generation of street life through café culture and an evening economy; place marketing; and the establishment of a resource centre to support arts community development. The formula that Montgomery has developed is derived predominantly from an analysis of British, Irish, and Australian central-city case studies of planned cultural and creative industry quarters. It is unlikely that such a carefully programmed approach to cultural quarter development can easily be transposed into a suburban North American context, where distances are significant, car-dependence high, nine-to-five commuting patterns common, and pedestrian foot traffic low.

Certainly elements of Montgomery's formula can be seen in the increasing popularity of public art and cultural events programming to the place-marketing campaigns of Canadian suburban municipalities.

To build a positive image of place, create opportunities for economic growth, and bring different stakeholders together, many Canadian cities use festivals and cultural events to promote themselves as distinctive places in the "experience economy" (Johansson & Kociatkiewicz, 2011). Engagement with the experience economy demands that cities convey an "attractive atmosphere, which comes from place-bound activities, events and services, attractive places and social spaces, which make visitors and residents feel inspired, involved, and connected to a place" (Lorentzen, 2009, 840). The City of Mississauga, for example, relies on free outdoor festivals such as My Mississauga or Celebrate the Summer at City Centre to unify, animate, and attract local residents to its suburban downtown with choreographed activities and shows. One such city-funded summer festival is a youth-oriented music event, Beating Heart Festival, held on the concrete square of the Mississauga Central Library and organized by a local promoter familiar with the local independent music scene. Although the promoter had organized several well-attended concert series that mixed hardcore heavy metal with electronic, acoustic, afrobeat, and fusion funk bands (which he characterized as musically accessible but not typical and generic), his vision for the festival was challenged by city officials. The municipality has established a festival mandate focused on attracting youth, yet they do not want to incorporate cutting-edge trends and alternative forms of music. In the words of the Beating Heart Festival promoter: "The city representatives let me know that they wanted the festival to be less 'rock show' and more accessible. It had to do with noise complaints they received from a few condo owners during some of our louder gigs" (arts administrator, Mississauga, 15 December 2007). Frustrated at the amount of time, energy, and commitment he has given to developing this festival while also holding down a full-time job, the organizer is not interested in musically branding Mississauga with what he describes as a politically palatable "radio rock band" concert series of "tame music." The formulaic approach to festival organizing that city managers were comfortable with did little to profile the distinctive music scene in Mississauga and, for this arts administrator, resulted in an event that lost its unique connection to a place (Quinn, 2005).

In 2008, 10 festivals or events in Mississauga drew crowds of over 5,000: MuslimFest (5,350); Saint Thyageraja Festival (6,050); Malton Community Festival (7,000); Mississauga Santa Claus Parade (30,000); Port Credit Canada Day (40,000); Streetsville Canada Day (40,000); Southside Shuffle Blues and Jazz Festival (75,000); Mississauga

Waterfront Festival (54,000); Carassauga (244,533); and Ribfest (90,000) (City of Mississauga Culture Division, 2009). Ironically, the "Mississauga/Toronto West Tourism Board believes that while festivals and attractions have strong appeal to local residents, none currently offers a unique enough tourism experience or attract enough participants to warrant investing in a marketing campaign beyond the city's borders" (City of Mississauga Culture Division, 2009, 8). Thus, while decision makers in Mississauga may consider festivals to be a municipal cultural strength and may write them into civic cultural policy and invest significant public resources there, the external perception of Mississauga's festival success is different.

In addition to an increase in cultural events planning, the desire for a cultural district is apparent in suburban cultural planning documents. Mississauga's cultural plan, for example, seeks to identify "nodes" (e.g., Cooksville, Clarkson Village, and Meadowvale) and "corridors" (e.g., downtown, Port Credit, and Streetsville) with concentrations of cultural resources (City of Mississauga Culture Division, 2009). Cultural nodes are also identified in the cultural plan as "places with an urban structure that is well suited to support cultural resources: walkable streets, an attractive public realm, a mix of uses that cater to a diversity of needs, and good access to public transit" (City of Mississauga Culture Division, 2009, 45). As well as concentrations of cultural resources in Mississauga's "urban villages" of Port Credit and Streetsville, the plan emphasizes the importance of the City Centre as a cultural node because of the presence of over 90 indoor and outdoor, small and large, informal and formal cultural spaces, as identified in the cultural resources database. Cultural planning advisers emphasize that Mississauga has a "limited supply of both dense, layered urban environments that creative communities tend to cluster in and industrial brick and beam properties that artists and creative entrepreneurs are drawn to as live or live/work space" (City of Mississauga Culture Division, 2009, 17). Benchmarks of urban culture are applied to the suburbs without alteration; unsurprisingly, the suburbs are found wanting. The challenge for Mississauga and other suburban contexts is perceived to be "how to nurture a critical mass of arts and cultural activity and facilities and a density of relationships and networks to support the evolution of a rich cultural ecology" like that which can be found in the central city (City of Mississauga Culture Division, 2009, 17). A common misperception is that a lack of a critical mass signals a reduction in cultural richness, which in turn motivates many suburban municipalities to move quickly to identify concentrations of cultural resources and activities.

Like Mississauga's cultural plan, North Vancouver's (Cultural Plan Steering Committee, 2002, 18) also outlines a strategy to identify locations "which would be suitable for development as cultural precincts." A cultural precinct in this document is explicitly understood as "a designated part of the community where arts flourish through specialized urban planning, which allows for a clustering of cultural facilities, programmed street activities, arts markets, and public art. All of which work together to create a vibrant arts hub that is attractive to citizens and tourists alike" (Cultural Plan Steering Committee, 2002, 18). Yet the concentrated diversity of cultural venues that deliberately support professional cultural production in an animated, intimately textured setting is precisely what is usually absent in the suburban context and is difficult to create from scratch.

One of the most commonly identified hurdles to suburban creative practice is that cultural infrastructure is neither as sufficiently developed nor as concentrated as in the central city. Scholars of the cultural economy repeatedly emphasize that concentration allows local cultural clusters to share many of the aggregation benefits of industrial districts; increased face-to-face contact augments opportunities for formal and informal social encounters, information and resource sharing, and collaboration. But underlying this list of features is the mistaken assumption that creativity occurs only when a critical mass of cultural practitioners and arts providers has visibly accumulated in the landscape. Creativity is equated with spatial concentration. Many Canadian suburban municipalities have proceeded to nurture creativity within their borders by trying to reap some of the agglomeration benefits of spatial proximity through investment in a single multi-purpose arts facility. But, as van Heur (2009, 1547 and 1546) has persuasively argued, through a detailed empirical study of the music industry in London and Berlin, "clusters are not spatial concentrations of creativity in any straightforward sense"; the cluster concept is just an over-used regulatory tool that policy makers have "latched on to … for transforming cultural production into an economically successful sector." Creativity does not need to be extroverted, visible, or concentrated to be authentic, transformative, or powerful.

Suburban Cultural Infrastructure and the Flagship Cultural Multiplex

"If 'downtown' once represented the central quarter of culture, it now faces stiff competition across the distributed city from purpose-built complexes in non-central locations" (Wright, 2008, 127). By way of example,

Herbert Wright cites the Paul Getty Museum in California, which opened in 1997 20 kilometres west of downtown Los Angeles. Richard Ingersol (2006) provides additional European examples of cities that have moved some of their new cultural institutions away from the city centre to the periphery.[15] In the American and European contexts funding and the political will exist to hire internationally renowned architects to design purpose-built cultural landmarks on the outskirts of some central cities.

It would be hard to argue in the Canadian context that downtown cultural facilities are under intense competition from suburban cultural or architectural landmarks. However, it would be fair to say that recognition is emerging in administrative, scholarly, and policymaking circles that the "growing cultural life of Canada's suburbs and edge cities requires investment in cultural infrastructure to ensure the vibrancy, development, and sustainability of these communities' unique cultural identities" (Duxbury, 2008, 10). Such investment needs to occur at various spatial scales (from the neighbourhood to the district to the city) and in multiple, accessible locales in order to begin to meet the various needs of a dispersed and culturally diverse population base. Over the last two decades, however, a significant portion of cultural infrastructure investment in Canadian suburban municipalities has been sunk into single, large, multi-purpose community cultural centres – the cultural multiplex.

All-in-one cultural buildings, which resemble a microcosm of a cultural district, are a new spin on the multiplex concept.[16] Cultural multiplexes are multi-use centres in which arts, culture, heritage, and library facilities may be housed under the same roof and share resources and operation costs. Rather than separate production from creation, and presentation from exhibition, cultural multiplexes may offer a continuum of these activities across a range of different artistic disciplines. Such cross-disciplinary usage and information exchange suggests that cultural multiplexes have the potential to be interpreted as miniature cultural districts.

In the Greater Toronto Area, the Peel Heritage Complex in Brampton and the Living Arts Centre in Mississauga are excellent contemporary examples of cultural multiplexes. The Peel Heritage Complex is a facility that contains the region's archives, an art gallery, a museum, the historic Peel County Jail building, and a community gallery, while also providing space for offices, workshops, and studios. In Metro Vancouver, a handful of multi-purpose community cultural gathering places have been built or revitalized over the last 20 years: the Richmond Cultural Centre;

Table 5.1   Employment in arts, culture, recreation, and sport in Surrey in 2006

	Cloverdale	Fleetwood	Guilford	Newton	South Surrey	Whalley	City Centre	Surrey
Labour force	21,765	27,545	27,865	58,880	30,180	4,695	9,705	210,930
Arts	375	520	585	1,030	1,010	675	195	4,205
% Arts	1.7%	1.8%	2.1%	1.7%	3.3%	13.4%	2.0%	2%

(Modified from a *City of Surrey Labour Force Fact Sheet*, City of Surrey Planning and Development Department, 2009a).

the Shadboldt Arts Centre (Burnaby),[17] the Port Moody Arts Centre and Inlet Theatre (Port Moody), and Evergreen Cultural Centre (Coquitlam). Significant renovations and expansion have occurred at the Surrey Arts Centre (Surrey) and the Leigh Centre Community Arts Village (Port Coquitlam). To reflect in greater detail the challenges and opportunities of suburban cultural infrastructure provision, two case studies of flagship cultural multiplexes are explored: the Surrey Arts Centre and the Living Arts Centre. The term flagship is used here to emphasize how the municipalities of Surrey and Mississauga have used these "talismanic," "'top-down' iconic projects" to render government cultural policy and action visible in the suburban landscape as building landmarks and cultural artefacts intended to bolster city image and attract further investment (Smith & Strand, 2011, 95 and 97).

The Surrey Arts Centre

Located in the community of Newton near the intersection of the King George Highway and 88 Avenue, at the virtual epicentre of the City of Surrey, the Surrey Arts Centre (SAC) has been municipally designated as "the focal point for the arts in the city." Ironically, the actual emerging City Centre of Surrey lacks an arts and heritage focus (www.surrey .ca). As the table below indicates numerically, Newton, where the SAC is located, and South Surrey are two of the communities in Surrey with the most people employed in arts, culture, recreation, and sport (Table 5.1).

The SAC is governed by the City of Surrey through the appointed Surrey Parks, Recreation, and Culture Commission, and the majority of the city's Arts Service Division experiences are generated and based here (PERC, 2008). This centralized facility specializes in hosting theatre, music, dance, and fine arts performances and workshops. The SAC

Figure 5.6   The Surrey Arts Centre (photograph by A. Bain).

contains exhibition halls, a 402-seat proscenium theatre with an orchestra pit and fly gallery, a 130-seat black box theatre with a sprung lino-
leum floor, a fully wired program room (that supports the production
and presentation of digital art forms), pottery and fine art studios, and a
gift shop that operates an art rental program (Figure 5.6).

The primary tenant of the SAC is the Surrey Art Gallery (SAG)
(Figure 5.7). As a civic art museum, the SAG is accountable to the municipality for its expenditures and funding and must be seen to be responsive
to the needs of the city in its programming. However, the management
partnership of culture and recreation presents challenges to curatorial
staff as the language, conceptual frameworks, operating methodologies,
values, and funding priorities employed in arts programming are often
quite different from those of recreation programming.

A further challenge to garnering funding support in Surrey is the
absence of additional commercial galleries or artist-run centres dedicated to the sale and exhibition of original art. The SAG is the only

Figure 5.7   The Surrey Art Gallery, The Art of the Sari exhibit (2007) (photograph by A. Bain).

venue in the city for professional art exhibitions. The curator jokingly remarked in our interview: "There's nothing else out there other than the community peg board in the local mall" (15 May 2007). The curators of Surrey's only professional exhibition space receive and review hundreds of unsolicited portfolios and project proposals a year. "Our limited resources present an ongoing challenge; although many projects and opportunities are proposed to us, we continually have to weigh and prioritize them within the bigger context of what's best for Surrey, for the gallery, and for the purposes of art and our audiences" (Davison, 2001, 24). Given that the SAG has had the same operating budget since 1995, it is amazing that a city with no grant programs for artists and an underfunded museum has been able to accomplish as much as it has.

Various exhibition catalogues published by the SAG document and celebrate the history and accomplishments of the museum (Davison 2001; 2002; 2005). In 1975, local citizens successfully lobbied city council to shift funding originally intended for a civic golf course into the construction of a "good gallery." This museum had a founding mandate to become an experimental space for leading western Canadian artists and to provide Surrey's many varied communities with access to the meaningfulness and relevance of contemporary art. Strong community support allowed for two expansions of the facility between 1975 and 1993, providing a new wing of exhibition halls that meet national museum standards. In 1995, when city council sought to eliminate the position of gallery director, the SAG made national news when community members protested with signs reading "Cut grass, not art." The Surrey city council responded to the protests with a promise to maintain the gallery's program budget.

The following year, the city created a new position, Manager of Arts Services, and embarked on a Cultural Strategic Plan. The new management structure and the adoption of the cultural plan by city council in 1999 signalled a renewed commitment to arts and culture on the part of city staff. A key component of the cultural plan was the redevelopment of the SAC, which included expanding the exhibition halls and studio classrooms and adding new space for a digital media lab and public programs. In 2002, a $7.5 million redevelopment of the SAC transformed the SAG into the third-largest public art gallery by floor area in British Columbia, after Vancouver and Victoria. Over a decade ago, it had already presented over 400 exhibitions, showcased the work of over 10,000 professional and amateur artists (e.g., local school children and community art club members), and attracted over a million child and adult visitors, predominantly from Surrey, Langley, and the Fraser Valley (Davison, 2001).

The success of the SAG since 1995 can be attributed largely to the hard work of its two curators. They have worked together "to be informed and influenced by the needs and issues of Surrey and its residents, and to respond with appropriate programming" that is intended to be designed *with*, not *for*, the community (Davison, 2001, 16). In the development of arts programming for the SAG, the curators seek to anticipate possibilities and provide opportunities out of which a sense of community pride and community capacity might grow. By being mindful of other perspectives and by valuing inclusive working relationships the curators seek to create and maintain a perception of the gallery as a safe and welcoming space in social, psychological, and spiritual terms. Over the years, they have extended invitations to many different ethnic and social groups in

Surrey to use the spaces and the resources of the gallery to voice their thoughts and experiences through art. A small sampling of these groups includes elementary and secondary schoolchildren, the Mayan refugee community, the Surrey Aboriginal Cultural Society, and the South Asian community. The curators hope that local residents will also see themselves reflected in the spaces of the arts centre. Thus visitors to the SAG are welcomed by sculptural banners in the parking lot that mirror the faces of the South Asian seniors who use the surrounding Bear Creek Park and totem poles from founding First Nations communities. This symbolism also carries through into the museum's community arts programming.

Community-based arts privilege process over product, and the SAG has built a national reputation based on its community arts programming. It aims to "build consensus among people with different perspectives, gifts, talents, and skills" (Kneight and Schwarzman, 2005, xii). By emphasizing "the right of excluded communities to assert their place in cultural life, to give expression to their own cultural values and histories" (Goldbard, 2006, 51), community-based arts strive to achieve understanding across differences and in the process provoke critical reflection, dialogue, participation, and action among multiple audiences (Kneight and Schwarzman, 2005). The SAG curators are sensitive to the changing demographics of Surrey, to topical social issues that require a safe space for expression, and to the need to reach out in innovative ways to the community. The SAG has helped to give a focal point and an identity to the community through entertaining and educating local residents.

Exhibitions and events held within the SAC gallery and performance spaces provide the city with a rich cultural program that has become both a civic and a tourist attraction. The cultural-space programming frequently provides opportunities to highlight the cultural wealth and potential of Surrey and to foster unity amid diversity. In this way, the SAC is a key symbolic space in Surrey: it has been designed and used by arts managers and (sub)urban decision makers to lend legitimacy to a narrative of urban growth and civic identity. As the single cultural centre for a dispersed community that has an underdeveloped not-for-profit sector, the SAC functions as a production hub, a meeting place, and a safe place of dialogue and knowledge exchange for professional cultural workers, amateur artists, arts organizations, and audiences.

Despite the SAC's remarkable degree of community outreach, there are practical limits to what one cultural facility can accomplish. As one

local visual artist (Surrey, 12 June 2007) who had the financial means
and initiative to establish his own art gallery and studio in a strip mall
explains:

> We have one gallery here, the Surrey Art Gallery. Suppose I want to have
> a show. We have to give a proposal and then take a decision and maybe in
> a couple of years I'll get a chance to show ... There should be some sort
> of commercial building where I can rent it for a week and display my work
> and nobody's using it. I can display whatever I like and people come and
> see it and buy. That sort of thing is missing here. Suppose I did not have
> this space to work and I am at home; I would need a studio place to work.
> There should be some space provided by the city. Even for the performing
> artists, they need space.

Neither the SAG nor the SAC is able to meet the very real, ever-present,
and practical demands for affordable work, exhibition, and performance
space by local amateur and professional arts practitioners. There are just
not enough days in the year and space within the SAG to accommodate
the needs of an expanding suburban cultural worker population.

The Living Arts Centre

On the other side of the country, the Living Arts Centre (LAC) is the
flagship cultural multiplex for Mississauga (Figure 5.8). Centrally located
within Mississauga's growing suburban downtown, the LAC sits on a six-
acre site across from the Square One shopping mall, just north of the
Civic Centre and the Central Library, and surrounded by condominium
towers. It was built in 1997 by Zeidler Partnership Architects (www.zgpa.
net) at a cost of $51 million. This 19,500-square-metre, multi-purpose
community arts facility contains two buildings, an arts and crafts building
and a performing arts building, that are linked via a large glass atrium
to appear as one.

The arts and crafts building has seven professional art studios (glass,
wood, textile, ceramics, drawing, metal, and photography), a small gallery with an annual rotating exhibition schedule mandated to profile the
work of local artists, a gift shop showcasing the work created on-site by
resident artists, and eight meeting rooms. The performing arts building
is composed of a restaurant and three theatres with professional lighting
and sound: Hammerson Hall, a 1,500-seat concert theatre; RBC Theatre,
a 400-seat recital theatre; and Rogers Theatre, a 110-seat, fully wired

Figure 5.8   The Living Arts Centre in Mississauga (photograph by L. Wheelans).

multi-purpose space. In the original plans for the LAC, the primary concert hall was to have 2,000 seats, but budget constraints reduced that
number to 1,500. Facility managers have since come to realize that this
decrease in the number of seats has affected the quality of touring acts
that can be brought in and the length of the tours. Many of the more
financially lucrative, higher-profile artists "won't come up here because
they consider it a suburb and they don't play in suburbs; they only play
in major cities" (arts administrator, Mississauga, 25 October 2007),
choosing larger downtown Toronto venues or the casinos in Orillia and
Niagara Falls over Mississauga. It remains a challenge to sell out the LAC
concert hall and to attract new audience members.

Like many cultural flagship institutions, the LAC is not financially
self-sufficient. The LAC is subsidized by the City of Mississauga with the
expectation that is should turn a profit for the municipality. While the
LAC receives the bulk of Mississauga's arts funding, it operates independently of the city, with its own non-unionized staff, chief executive offi-
cer, and board of directors. Although it was intended as a centrepiece for

culture in Mississauga, this flagship cultural multiplex is underused and has very little drawing power.

A frequent refrain from Mississauga's cultural workers is frustration with the unfulfilled potential of this architecturally appealing community arts facility. As one photographer (Mississauga, 11 November 2007) explains: "You have a place like the Living Arts Centre but that's not going to pull people in. It doesn't even pull people from Mississauga in, let alone further afield. Every time you're in there, there's not much happening, which is a real shame, because it's a beautiful facility." A similar experience is described by a local sculptor (Mississauga, 18 October 2006) who once rented studio space part-time: "If you walk into the Living Arts Centre during the day, the place is empty and it's dead. There's nothing, unless they're filming a movie and the only creative activity is the people in the studios." Both of these cultural workers characterize this flagship building as a cultural void rather than as a cultural hub. For many members of the local cultural community the space possesses the uninviting feeling of a corporate building. Although a handful of studios and the offices of the Mississauga Symphony Orchestra are occupied, this is not enough creative activity to animate the entire building throughout the week. Part of the reason for the absence of cultural activity on weekdays may be the financial inaccessibility of the facilities to individual practitioners, cultural organizations, and community groups.[18] Out of necessity, grant money issued by the Mississauga Arts Council (on behalf of the City of Mississauga) to arts groups, who may be tenants of the LAC, often goes directly back to the city through rent for performance or studio space. Some non-profit groups that organize community events have benefited from the negotiation of a temporary reduction in user fees, as the managers of the LAC seek to reach out to new audiences.[19]

The primary users of the LAC are the business community. Rather than profile the work of resident artists, or the studio rental rates, or the studio layouts and equipment, the LAC's website (www.livingartscentre.ca) is geared to marketing high-profile performers in its largest theatre and packaging the venue as a rental facility with catering for all occasions. Thus it is easier to find out that the atrium can host 350 guests and has been used successfully for trade shows, production demonstrations, corporate dinners, weddings, and parties than it is to discover what group exhibition is currently on at the gallery. Complaints abound that the LAC isn't "an audience-friendly place where people go out for supper and then come to a performance" (conductor, Mississauga, 11 October 2006). The restaurant remains empty most of the time and the food is frequently described as being of poor quality. While the restaurant may be empty, the two levels of below-grade free parking for 450 cars at the LAC are full – not with artists, but with visiting employees of corporations (e.g., Walmart) and organizations (e.g., Ontario Teacher's Association) which can afford to rent the entire building. Unlike the community-building success of the SAC, what could have been an active cultural gathering place for artists has become a corporate meeting ground instead.

Conclusion: More than Suburban Cultural Boosterism

Together, the Surrey Arts Centre and the Living Arts Centre offer an interesting perspective on the role of cultural multiplexes as development boosters in young suburban municipalities. A conventional dictionary interpretation of the verb "to boost" is to increase the reputation or value of a person, a scheme, or a commodity through advertising. *Boosterism* has a more precise meaning. In nineteenth-century North America, boosterism referred to the act of promoting a city to encourage growth (Short, 1999). The conservative doctrine of growth was espoused by local politicians, business élites, property owners, real estate agents, and the media through promotional activities such as exhibitions, films, ceremonies, advertising, and publications (Huskins, 1999). "Booster actions were depicted as natural, indeed inevitable, products of a single historical trend called 'progress'" (Tronrud, 1993, 49). During the expansion of western Canada and the United States, local boosters needed to persuade others to share in their dreams and invest in their community if it was to survive (Gruneau and Whitson, 1993). "Whether an area would develop as an important urban centre or 'turn to dust' depended very much on the ability of its civic leaders to attract industry, jobs, and attractive urban recreation for its citizens. In this context, it was necessary to cultivate an image of prosperity, of the area's entrepreneurial drive and boldness of vision" (Lowes, 2002, 17). Contemporary boosterists are engaged with advertising their city as a place to do business and as an interesting place to live and to visit (Newman, 1999). As discussed in earlier chapters, culture has been given an increasingly prominent role in competitive civic place-making strategies. Civic élites in the suburban municipalities of Surrey and Mississauga have bought into the idea that flagship cultural facilities and festivals can uniquely brand their cities and project an image of cultural sophistication and economic dynamism, much like their central-city counterparts in Vancouver and Toronto. The difference is that these suburban municipalities have each relied upon one high-profile, centrally located, municipally funded community cultural centre to do the work of several urban cultural districts composed of differently sized venues, supporting artistic production and consumption across a skill spectrum from amateur to professional.

In keeping with the creative capital model of urban development, the SAC and the LAC are cultural spectacles that have been made central to the re-imaging of their respective suburban municipalities. Both cultural megaprojects are located close to their civic-centre funders and decision makers at the symbolic physical and political centre of their municipalities. While they may lack an internationally renowned architect as building author, these prestige suburban cultural edifices, with their branded spaces of performance, presentation, and creation, were designed to be bold visual statements on the suburban landscape. Dramatic in their form, building material, and use of colour, light, and texture, the interior and exterior spaces of the SAC and the LAC were intended to be distinctive, to capture attention, and to photograph well. With their high profile in city websites and brochures, their prominent marking on electronic and papers maps, and their visible directional street signage, the SAC and the LAC have become the dominant face of culture for their respective municipalities. But such use of cultural facilities for place marketing invariably distorts what they are able to do and whom they are able to serve on the ground. Thus these "literal 'construction[s] of culture'" are rendered "poor substitute[s] for a city's genuine cultural development" (Smith & Strand, 2011, 97). Regardless of the extent of marketing, public investment, and political expectations, it remains a real challenge for any one building to effectively meet all of the cultural infrastructure needs of a diverse and dispersed community that includes professionals, semi-professionals, and amateurs, established and emerging cultural workers and organizations, as well as newcomers and long-term residents.

With their large populations, Surrey and Mississauga have remarkably underdeveloped hard cultural infrastructure. What cultural assets these cities do possess are not evenly distributed across the municipality within the different town and population centres. In Surrey, as innovative and successful as the SAC and the SAG are in their arts programming and community outreach, there are inevitable gaps in cultural service provision. Through letters and cultural planning surveys, the Arts Council of Surrey and its many member organizations and individual artists have made city officials aware of the shortage of adequate work, exhibition, and performance space in specialized cultural facilities. The following letter sent to a Surrey artist (13 June 2007) from the Surrey Arts Manager is a typical bureaucratic response to a request for more space:

Every member of the senior management for Surrey Parks, Recreation and Culture Department is aware of your group's need for a pottery studio space. We have explored the issue with the real estate division in terms of all city properties and there is nothing suitable or available at this time. I will certainly let you know, if I hear of any possible solutions on either public or private property.

The lack of purpose-built space to support cultural production and cultural consumption has been duly noted by Surrey city officials, yet no public investment commitment has been made to address the shortfall of cultural facilities. In the interim, artists and arts groups continue to make use of multi-purpose spaces scattered throughout the municipality: residential basements and garages, and – when they can afford the rental rates – banquet halls, community centres, libraries, churches, and high school theatres. Such alternative, multi-purpose venues can be valuable temporary cultural anchors that help to nurture a germinal local artistic scene, but they get little of the municipal resources and media attention given to the flagship centres.

In the suburbs, as in the central city, culture in its consumptive and spectacular form has the upper hand over production and practice. In terms of political funding and public investment, the requests of prestige cultural multiplexes and the festivals that help to animate them get priority over smaller, grassroots arts organizations that continue to struggle to obtain funding to upgrade their spaces and/or to keep them viable. Such a cultural branding strategy may enhance the symbolic value of a city, but it undermines the conditions of everyday work that are necessary to sustain the creative process (Bader & Scharenberg, 2010). It remains challenging for many Canadian arts and culture organizations to obtain adequate capital to fund their own facilities. Yet such funding is essential: the smaller arts organizations and facilities often help to form the matrix that underpins and sustains the larger cultural economy of a city-region. In light of the cultural infrastructure shortages experienced by many suburban municipalities with ethnically diverse and spatially dispersed populations, a potentially more socially and economically sustainable approach to cultural development would be to value the seemingly unspectacular and informal community service hubs that can function as suburban cultural magnets. When cultural policy and investment are directed into smaller-scale, multi-purpose arts facilities and the flexible networks that link and animate these spaces, there is greater potential to engage a wider audience for the role of culture in building communities and places.

Chapter Six

The Creative Art of Place Complexity in Suburbia

The city is an extremely complicated thing. Of all the [hu]man-made things we've made – flying machines, floating machines, digging machines – the city machine is by far the most complicated modern construction. When you start multiplying all the possibilities of what can happen where, and how, and in what order, the city becomes this uncontrollable, complex place.

— (Artist and architect Michael Awad cited in Mays, 2008, G2)

Creativity comes in many different forms and is stimulated by many different environments, not all of which are large and mature urban centres with branded arts districts. Sir Peter Hall (1998) has long argued that creative processes and cultural innovation are often concentrated in declining places. In his book *Cities in Civilization*, Hall (1998, 286) writes that creative cities "are places of great social and intellectual turbulence, not comfortable places at all." Hall (1998, 288) claims that "creative cities are almost certainly uncomfortable, unstable cities, cities in some kind of basic collective self-examination, cities in the course of kicking over their traces." While Hall has deliberately chosen to use the metropolitan signifier "city" rather than "suburb," his assertion of the value of discomfort, uncertainty, and scrutiny to the creative process would certainly apply to the place-making currently unfolding in suburbia. Areas within North American suburbs can also be considered creative. To outsiders, they may not be as identifiable as their urban (milieux, quarters, and districts) counterparts, and they may also vary in terms of scale, form, density, and usage, but they nevertheless deserve the descriptive label "creative."

In his book *The Marriage Benefit* (2008), the clinical psychologist and psychotherapist Mark O'Connell describes marriage "as a forum for learning through a kind of creative constraint" (O'Connell, cited in Fillion, 2008, 30). The same idea could be applied to the suburbs: a place where cultural workers can learn to find meaning precisely because there are limitations. This chapter examines how the very infrastructural constraints of undertaking cultural work in the ambiguous, in-between landscapes of the suburbs can sometimes encourage greater innovation and collaboration between cultural workers and local residents. Particular attention is directed to the establishment of informal, multi-purpose community cultural service hubs in retrofitted storefronts, strip malls, and schools. Case studies of grassroots cultural economy interventions show how these cultural service hubs, maintained often through voluntary reciprocity, act as creativity incubators, productively recombining people and social groups to create new collaborative opportunities, spin-off activities, and unexpected synergies. In addition to possessing a gentle, gravity-like pull for suburban creatives, these unofficial service hubs add valuable layers of under-appreciated complexity to suburban places. The focus of this chapter is on cultural service hubs as suburban creative milieux and on the role of cultural workers as suburban place-makers.

A Love-Hate Relationship with Suburbia

It seems as if everyone has an opinion on the suburbs (Beuka, 2004). One could go so far as to say that everybody "loves to hate the suburbs, except for the people who live there, people, who, sometime in the last few years, became a majority" (Peterson 1999, 45 cited in Kolb, 2008, 144). This quotation underscores a long-standing tension in the attitudes toward suburbia held by professionals and academics versus ordinary suburban residents. While suburbanites might consistently praise their residential environments for meeting the necessities of everyday family and community life, urbanists have repeatedly wielded a long list of pejoratives to describe the suburbs: kitschy, crass, thin, diluted, simplified, tame, sterile, barren, bland, drab, stark, mediocre, prefabricated, uniform, sanitized, conforming, alienating, consumerist, self-serving, undistinguished, and unmemorable.

Using British pop music lyrics and imagery as his starting point, the cultural critic Simon Frith (1997, 275–6) offers a cynical riff on the basic components of suburbia:

> Suburban dwellers move there from somewhere else; suburban housing is designed for the small family; suburban lives are lived behind closed doors. There's no sense of excess here; no spill-over of cousins, aunts and uncles;

no massing on street corners. There are single-class communities: people don't know each other but they know what they're like. Neighbours nod across the street, compare cars, keep their salaries to themselves. Suburbia is a place where people live but don't work; rest but don't play (the real jobs, the real shops, the real pleasures, are elsewhere). Geographically, suburbia is, in effect, an empty sign, a series of dots on the map from which people travel – to the office, to the fleshpots, to the city. Suburban living is characterized by what it lacks – culture, variety, surprise – not by what it offers – safety, privacy, convenience.

Frith notes, however, that this description of suburbia is fantastical. Suburbs are no more classless, genderless, cultureless, or placeless than anywhere else. The derision, he asserts, allows scholars, urban planners, politicians, journalists, and media commentators to dismiss suburbs as "homogenized, soulless, plastic landscape[s] of tepid conformity, an alienating 'no place'" (Beuka, 2004, 4). Suburbs are thus positioned as inferior to cities in social, cultural, economic, political, and physical terms, and their authenticity, together with anything produced there, is called into question (Martinson, 2000). To talk of suburbia, Barry Langford (2000, 68) writes, "instantly invites a certain jargon of authenticity – authenticity, that is, being what the lonely crowd of the suburbs, construed as secondhand, retentive, hypocritical, conformist, soulless, definitively lack." Thus suburbia is labelled as a creative backwater from which nothing of genuine cultural value or originality can materialize.

When the suburbs are repeatedly scapegoated and "othered" (e.g., for their dependence on the automobile, their destruction of nature, increased consumerism, uniformity, and inauthenticity), it becomes a challenge for those living there to find opportunities for imagining and enacting alternative presents and futures, and to open up those possibilities to public scrutiny and debate. This chapter shows, however, that in many suburbs it is cultural workers who are at the forefront of challenges to, and reconfigurations of, the (mis)perceived banality of suburbia. It is cultural workers who revel in the place-making possibilities of a not yet fully formed locale, like this mixed media artist from North Vancouver (16 May 2007): "It's really lovely and makes it so exciting and so interesting that you're in a place that is not mature yet, and you are part of the process." A photographer from Mississauga (11 November 2007) elaborates on the challenges and the possibilities of suburbia:

The work I'm doing right now, as an artist, you have to look harder and think much harder in the suburbs. You have to really look for things. They don't really come and jump out at you. But then it can be very inspiring too, and it is possible to live in a cookie-cutter house and drive an SUV and be an artist and be creative. You just have to work that much harder at it and – you know what? – that's not a bad thing.

Through their creative labours cultural workers can uncover the "complex internal multiplicities" of suburbs to reveal how the suburbs are uniquely textured with their own intensities and elasticities (Kolb, 2008, 103).

Sense of Place, Identity, and Complexity

The geographer Ted Relph (1976, 90) was one of the earliest scholars to critically examine placelessness as a dominant cultural force in urban settings, a force said to weaken the distinct identities of places to the "point where they not only look alike but feel alike and offer the same possibilities for experience." Relph criticizes modernist suburban landscapes as materializations of scientific rationality and spatial efficiency in the service of profit. For Relph, suburbs epitomize placeless locales. With their "instantly identifiable, uniform architectural styles and landscape designs" suburbs have, over the last half-century, "contributed toward a proliferating sense of placelessness and in turn the perceived homogenization of American life" (Beuka, 2004, 2). Building on Relph's work, Paul Knox (2008, 34) writes that a "positive and distinctive sense of place stems in large part from routine encounters and shared experiences that make for intersubjectivity. This requires plenty of opportunities for informal, casual meetings and gossip; friendly bars and pubs and a variety of settings in which to purchase and/or consume food; street markets; a variety of comfortable places to sit, wait, and people watch; a sense of ease with changing seasons; and, above all, a sense of belonging, affection, hospitality, vitality, and historical continuity: not the attributes that anyone associates with American suburbia." For Knox, the social and the socializing features built into spaces matter because it is through human interactions that spaces are gradually transformed into more meaningful places to which people develop an emotional attachment. When people live their lives in places, they begin to feel a part of those places, rendering places meaningful.

How a space is used and made meaningful hinges on notions of legibility and imageability, urban planner Kevin Lynch (1962) has argued.[1] A city that possesses a compact built form with a legible "architectural hierarchy of monuments and urban spaces" is said to have a more readily recognizable identity than a sprawling suburb that lacks a "sense of syntax" and can be neither "read as a unity" nor grasped as a whole (Ingersol, 2006, 10). When the identity of a place is poorly defined, both to residents and to outsiders, it can be harder for people to build community attachments to that locale. Many of the cultural workers interviewed more easily identified themselves with the larger census metropolitan area within their city-region than with the suburb within which they live and work. One public art sculptor (Mississauga, 18 November 2006), for example, admits: "When I travel and am asked where I am from, I say Toronto, because it's a major city in Canada and it's close to where I'm from. But, if I were to say Mississauga, which I wouldn't because I'm not proud of being from here, they wouldn't know where it is." The absence of place pride stems, in part, from frustration with the perceived lack of recognition and support that this cultural worker has attained from his suburban municipality for his artwork and career achievements, and from the lack of visual inspiration that he finds in the suburban landscape. By way of comparison, a younger visual artist from Mississauga (3 November 2007) believes that more local cultural workers need to confidently assert their suburban lineage:

> I think that since artists and people who are really innovative abandon Mississauga, it's wrong. More people need to say: "I'm from Mississauga and look at what I'm doing. I'm not from Toronto I'm from Mississauga. I came from Mississauga and I'm doing something that's really interesting and I grew up in the suburbs." It doesn't mean that things have to be boring. I think that if more people from the suburbs claim that they are from the suburbs, people might look at them differently.

This quotation reinforces the importance to suburban municipal decision makers of not only retaining cultural talent but also of recognizing and encouraging it. There is collective persuasiveness in multiple individuals identifying with and proclaiming their allegiance to a place. Mississauga is not alone in its place-identity struggles to have a positive meaning willingly ascribed to it by residents. Many suburban communities in North America are "searching for a unifying identity" in which their residents articulate "a longing for a stronger sense of place, including stronger connections with people, local traditions, and nature" (Rodriguez, 1999, 2).

The notion that suburbs are not visually barren, homogeneous places is a message that has been asserted in the scholarly literature consistently over the last decade (e.g., Martinson, 2000; Whitehand & Carr, 2001; Hayden, 2003; Hudnut, 2003; Lucy & Phillips, 2003; Stanilov & Scheer, 2004; Kruse & Sugrue, 2006; Fiedler and Addie, 2008). The physical and social diversities of American suburbia have led Dolores Hayden (2003, 11) to argue that "in the spaces of the suburban city lie metropolitan complexities" that are both revealed and concealed. It is the role of culture and cultural workers in enhancing the complexities of suburban places that this chapter explores.

In his book *Sprawling Places*, David Kolb (2008) introduces the concept of complexity as an important normative criterion for understanding and evaluating contemporary places. The more common criterion of place authenticity, he suggests, is vague and overly judgmental of places that are not centred and hierarchical. Complexity, on the other hand, values the many ways in which places are constituted by "interacting multiplicities": "multiple roles, forces, norms, processes, internal spatial divisions and external links to other kinds of places and to the processes that bring together multiple forces and systems" (Kolb, 2008, 54). For Kolb, there are three dimensions to place complexity. First is structural complexity in the social norms and spatial arrangements of a place. Second is local interpretative complexity in the reproduction of the normative social and spatial structures of a place. A third dimension is participatory awareness complexity in the appreciation of the systemic political, economic, and environmental processes that shape a place.

Kolb describes structurally complex places as those that support a continuum of presences and absences, a non-linear path of action and interpretation, and a potential for conflicting internal tensions. "Complex places restrict what should follow upon what, while increasing the number of connections and normative possibilities within the place" (Kolb, 2008, 59). He suggests that "discontinuity and possibility imply potential complexity, since possibility and meaning cannot be completely controlled" (Kolb, 2008, 79). However, as Kolb hastens to add, complexity does not necessarily equate with the capacity to surprise or to reveal new information; nor does it necessarily produce a balanced and pleasing whole. Complexity and coherence, then, may not go hand in hand to contribute to the positive image of place that Lynch (1962) imagined. For Kolb (2008, 67), complexity becomes less about producing an interrelated totality and more about producing a "web of possibilities" within a place. When the complexity of a place is increased, Kolb (2008, 104) maintains that there is necessarily "less isolation, more connection and mediation, and more intertwined self-relations."

In the penultimate chapter of his book, Kolb (2008, 162) asks if "suburbs can become more self-aware about their own growing complexity and involvement in larger linkages and processes." Where Kolb turns to the planning paradigm of New Urbanism to answer this question, this book turns instead to the labour and products of cultural workers. The creative practices of cultural workers, as the rest of this chapter explores, play a dynamic and underappreciated role in fostering all three dimensions of complexity that Kolb identifies: spatial complexity, interpretative complexity, and participatory awareness complexity.

Hubs of Culture

A standard criticism of postwar suburbs is that they are fragmented into disparate sites of socialization and rarely possess physical, public centres in which people can convene (Martinson, 2000; Chiras & Wann, 2003). The home has traditionally been at the centre of suburban life.[2] On the surface, contemporary suburban social life appears to perpetuate a focus on the enclosed spaces of private homes, yards, and automobiles, and on privatized locations such as schools and shopping malls (Baumgartner, 1988). If, however, "you are shallow enough to judge a place for its face value, which is just the way it looks, which is just a Walmart (and) a bunch of homes, then the suburbs may appear boring. But if you look into every one of those homes, there is something unique happening … I think the suburbs [are] rich in culture and this goes with any city, you just have to look a little harder in the suburbs to find it. The suburbs [are] just short on venues to showcase the interesting and unique work [they have] but there is so much going on" (Marsella cited in Kobayashi, 2007). Inspired by the words of this 31-year-old founder and director of the Brampton Indie Arts Festival as quoted in a *Toronto Star* newspaper article titled "Toronto Looks to the 905 for Culture," the remainder of this chapter explores the place-making significance of informal community service hubs and creativity incubators established by suburban cultural workers.

The language of "hubs" and "incubators" has become a trendy way to describe cultural centres in the creative city. It has found its way into social policy documents, planning reports, and the news media.[3] In Toronto's policy and planning reports, the word hub is most commonly paired with creativity.

In Toronto's *Agenda for Prosperity* (2008a), creativity is highlighted as one of the four pillars in the mayor's strategic development framework.

In the action plan for "Creative Toronto" specific mention is made of the need to "adopt a cultural planning model to identify clusters of creative activity, evaluate cultural assets, promote a place-based approach to creative sector development and establish creative hubs and districts as geographical concentrations of interconnected individuals, organizations, and institutions involved in the arts, cultural industries, new media, design, knowledge-building and/or other creative sector pursuits" (City of Toronto, 2008a, 42). Within the context of this document, "creative hubs" are understood as "innovative platform[s] for combining the necessary hard and soft infrastructure to support the space and programming needs of commercial, not-for-profit, and community sectors (City of Toronto, 2008a, 42). It is, however, the needs of the commercial sector that are given priority. A supporting report prepared by the urban cultural policy consultants Greg Baeker and Glen Murray explains that creative hubs

> form from groupings of interconnected and interdependent businesses, institutions, places and scientific and cultural resources. When sufficiently networked, the creative activities of these various businesses, artists, scientists, and entrepreneurs converge to produce new ideas, products, services, art and design. They can form spontaneously, as a result of unplanned stimulus, or can be formally developed as convergence centres or platforms to integrate, accelerate and commercialize research. (City of Toronto, 2008b, 25)

This quotation illustrates how the concept of hub is used within cultural policy documentation in a technical and economistic way to assert the importance of places that function as infrastructure nodes and innovation platforms in the knowledge economy. The main beneficiaries and gatekeepers of such creative hubs are the same élite urban actors who benefit from growth coalitions and hold the reigns of decision-making power in a city – developers, landowners, businesses, business improvement associations, institutions, and local politicians (Catungal & Leslie, 2009a). Individual cultural practitioners often remain excluded from this coalition of the privileged. Through an analysis of Liberty Village, a 45-acre brownfield culture-led redevelopment site located just west of downtown Toronto, Catungal and Leslie (2009a) reveal how this creative hub advances the interests of large creative firms and commercial property developers. Although they neglect to define their use of the term hub, one can extrapolate from their discussion that they understand it to be a central place where creative production networks find spatial expression in the interests of profit.

In her analysis of Canadian cultural infrastructure Nancy Duxbury (2008) employs a metaphor of collision. She asserts that in Canada "incubator, convergence centre, and multi-sector convergence projects" extend a "co-location strategy to intentionally provide shared platforms and support for fostering creative experimentation, networking, and cross-sectoral 'collisions' and connections" (Duxbury, 2008, 82). In her estimation, formal cultural "convergence centres are vibrant physical places that are designed to maximize socialization, networking, and random collisions, and thus become major connecting hubs and economic engines" (Duxbury, 2008, 83). This chapter is concerned with a less commercially exploited and heavily marketed understanding of the hub concept within the cultural life of suburban communities. Instead, I am inspired by two reports commissioned by the Toronto Arts Foundation (2009; Resonance Creative Consulting Partners 2008) on neighbourhood arts hubs that initiated, in 2011, a new Neighbourhood Arts Network "dedicated to supporting art-making in neighbourhoods throughout the city" (www .neighbourhoodartsnetwork.org). This 500-member network has as its central resource a website maintained by a contract staff member at the Toronto Arts Foundation whose job is to organize monthly meetings that bring community-engaged artists together to share information and to stay connected. In a report, *A City Creative to its Bones* (2008), the consultants describe neighbourhood arts hubs as centres of people and energy from which creativity radiates. These hubs can be born of necessity or by accident, emerging organically to respond to a community's needs and interests. They link people who live in a neighbourhood to professional arts organizations and artists who work in a variety of media.

In Toronto, as part of a cultural mapping initiative, a report titled *A Map of Toronto's Cultural Facilities: A Cultural Facilities Analysis* (2003) makes a useful distinction between hub and incubator spaces. To reiterate, "hubs" are said to be non-traditional performance and exhibition venues that provide community-driven cultural activities, while "incubators" are said to be artist-run facilities that function as "a hothouse for the creation of new ideas, approaches and productions" (ERA Architects Inc. et al., 2003, 7). These definitions bring to the fore the labour of cultural workers and assert the importance of a community rather than a business agenda. As discussed in the previous chapter, the majority of formally identified hub and incubator spaces are reported to be located

Figure 6.1   The Neilson Park Creative Centre (photograph by L. Wheelans).

in "urban neighbourhoods" in downtown Toronto. Two of the few suburban cultural spaces in Toronto that are profiled in this report and on the city website[4] are the Neilson Park Creative Centre and the Assembly Hall in the inner suburb of Etobicoke. Each of these hubs will be introduced below and the kinds of creative cultural practices they support briefly discussed in order to provide a contrast to short case studies of the more grassroots community cultural hubs that go largely unacknowledged in cultural planning policy documents.

The Neilson Park Creative Centre

The Neilson Park Creative Centre was built in central Etobicoke in 1992 by the City of Etobicoke and is now owned by the City of Toronto. At the south end of a park, on what was formerly a tract of farmland, this community arts facility is in the middle of a wealthy residential subdivision and is largely inaccessible by public transit (Figure 6.1). The building was constructed in response to lobbying by five local amateur crafts groups for a space to create and to display their art. In exchange for the building, which is leased from the city for \$1 per year, the resident groups of painters, quilters, rug crafters, and spinners who manage the facility through an umbrella organization are required to promote the appreciation and study of the visual arts to the wider community. Nevertheless, since its inception, the centre has predominantly been run as a private members club that offers only the bare minimum to the surrounding community in the form of classes and potential rental facilities for art exhibits and workshops. Its two gallery and workshop spaces are heavily used by its 500 members, who are predominantly white, upper-middle-class retired women or stay-at-home mothers living in Etobicoke and Mississauga and who would describe themselves as art enthusiasts and hobbyists rather than professionals. The member groups rent studio space and gallery space, and what remains is rented out on a subjective, sliding scale to individuals and organizations. The bulk of the facility rental income is generated from companies who pay full price to hold business meetings and training workshops. The two gallery spaces are rented in three-week blocks, two years in advance, at a rate of \$750 for the main space and \$375 for the smaller hall space plus 10 per cent of sales. Of the 34 shows a year, only two are open to the public as juried shows with the goal of helping artists who might otherwise not have gallery representation or be able to afford to rent gallery space. The centre does minimal publicity, with occasional advertising in Toronto's *Slate Art Guide* (www.slateartguide.com). It has only recently begun to apply for arts grants.

In the words of the Neilson Park Creative Centre manager (8 November 2006): "We can and we should have a much, much bigger influence in the total of our community. We don't get involved with other art organizations ... We've been a little stagnant. It's our own little baby and we should diversify ourselves, get our hands wet, and our feet wet in all kinds of different things." The word "creative" is explicitly used in the name of the building. Yet when the manger was asked how creativity is interpreted in the kinds of programming or in the mandate of the facility, she could say only, "we have endless discussions about what is a craft versus what is art." In cultural planning reports the Neilson Park Creative Centre is officially designated as both a "hub" and an "incubator" space. However, the above quotation quite clearly conveys that this arts facility serves a select group of amateurs and does little to actively reach outside of this bubble of privilege to nurture artistic innovation and conversations with the wider cultural community.

Figure 6.2   The Assembly Hall (photograph by L. Wheelans).

The Assembly Hall

South of the Neilson Park Creative Centre, near the shores of Lake Ontario and the campus of Humber College, is the Assembly Hall (Figure 6.2). This is a century-old, city-run heritage building that was renovated at a cost of $4.5 million and reopened in 2001 as a multi-purpose community cultural centre. The manager of the Assembly Hall explains (13 November 2006) that the mandate of this facility is to function as "a gathering place at the heart of the Lakeshore where community and creativity are celebrated and nurtured." When asked how creativity is understood in this mission statement, she states: "Creativity is anything that's somehow related to arts ... where either people are coming to see other people's creative efforts or are coming to do something that is creative themselves." This is a rather vague interpretation of creativity, but one that perhaps accurately reflects the range of uses supported in this building. Intended mostly for theatre, the performance hall in the Assembly Hall seats 250 and is rented out on a sliding-fee scale to community theatre groups, to Humber College for student performances, and to individuals for private events. While there is a stage, there are only two small dressing rooms. The acoustics in the performance hall are not ideal for music. Fine art is displayed in the leftover spaces of the hallway and foyer, spaces that get very little people traffic unless a performance or meeting is scheduled. With over 300 permits issued a year, the facility hosts everything from cake decorating to children's drama classes, and it is not used strictly for arts programming. "Because it is one historic building that was carved up, almost everything is multi-purpose, multi-usage. It's a wedding today, it's a performance tomorrow, it's a community meeting the next day. There's not full amenities for anyone" (arts administrator, Etobicoke, 13 November 2006). This alternation of activities, the manager believes, will bring culture to a wider audience who might otherwise not cross the entrance threshold.

The Assembly Hall is classified in Toronto's cultural infrastructure planning documents as a "hub." As an adapted heritage property with a performance hall that is widely used by community groups, it certainly qualifies as a non-traditional performance and exhibition venue that supports local cultural activities. Culture, in this instance, is broadly interpreted to include a range of social and business events that are regarded by administrators as useful revenue generators to help cover maintenance and operating costs. Some might suggest that the cultural mandate and the cultural exchange capacity of the facility are diluted. Hubs are meant to function as channels of mediation between different social groups, but at the Assembly Hall activities occur and groups meet separately – they alternate rather than overlap. Given the reduced capacity of the Assembly Hall as a socially interactive space, its official labelling as a "hub" is misleading.

There are, however, two facilities in Etobicoke that are not formally documented in Toronto's cultural infrastructure mapping exercise but that do function under the cultural policy radar as invaluable community cultural hubs. They are the Waterfront Trail Artists Association and the Franklin Horner Community Centre.

The Waterfront Trail Artists Association

The Waterfront Trail Artists Association (WTAA) is a non-profit, artist-run organization that began in 1990 as splinter group of professional artists from the Lakeshore Arts Council who had concerns that the new arts council administration exploited their creative labour without adequate financial reimbursement and provided minimal support for their community arts projects. For much of the first decade, the WTAA operated out of the house of the main organizer, a 30-year resident of Etobicoke, running events in donated facilities and at free outdoor venues (e.g., parks, beaches, and streets). From 2006 until 2011 the WTAA had a stable residence in a former pool hall in a retail strip along Lakeshore Boulevard that was rented for $1,500 a month (figures 6.3a and 6.3b). The 330-square-metre open-concept space was informally divided with donated furniture and masking tape lines on the floor into a gallery, an office, a lounge, a classroom, a flexible art production space, and studios. The studios were intended to be afford-able and accessible, revealing the process of artistic production to the public who attended classes. These studios were rented out at a rate of $190 per month to members, with a $100 discount provided to those art-ists who volunteered at events or provided labour in kind. For example, one professional photographer ran local photography excursions and used the classroom space for group critiques; in return for free space he photographed WTAA events. In 2008, the organization ceased to obtain sustainable grant funding and by 2011 had fallen short on operating costs while trying to keep programming and events affordable. A cash-and-carry contents sale was held in support of the membership and the emptied space was taken over by a glass retailer. The WTAA remains a viable volunteer organization with a board of directors but no paid staff; members continue to run special events and educational programming in the community when they have time to do so. Without the physical space of the storefront, a Facebook page is now the organization's pri-mary point of contact.

Before the closure of the storefront, the WTAA had 80 members, 60 professional artists, and 20 art supporters and families, who each paid a membership fee of $5 a year and received an electronic newsletter detailing upcoming activities and classes for all ages (e.g., stained glass, watercolour, acrylic painting, life drawing, and animation). The intent of the founders was to create for local community members "a safe place to come to make art and learn about art, and talk about art" (WTAA board member, Etobicoke, 4 October 2006). The WTAA storefront became an intimate and experimental space where amateurs and professionals could learn from one another in an unintimidating setting and in a hands-on way. The storefront and the consistent track record of community arts events helped the WTAA demonstrate that they are more than an artists' collective and were for several years successful in securing some program funding from the Toronto Arts Council, the Ontario Arts Council, and the Ontario Trillium Foundation.

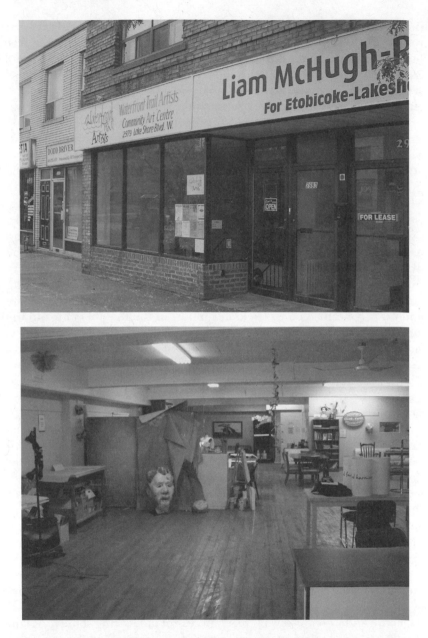

Figures 6.3a and 6.3b   The Waterfront Trail Artists Association renovated store-front and shared studios closed in the spring of 2011 (photographs by A. Bain).

From the outset, the WTAA has worked closely with community partners to demystify artistic labour and to reveal the hidden histories and geographies of south Etobicoke through event programming. One of the first year-long public art projects that WTAA organized for the millennium resulted in the creation of 2,000 papier-mâché replicas of now extinct indigenous birds (after which the town of Mimico was named) by elementary schoolchildren at in-school, artist-led workshops. In addition to bringing together artists, students, seniors, and environmental groups, the Flight of the Passenger Pigeons project increased environmental historical awareness by displaying this ornithological art in stores, libraries, and community centres.

WTAA's programming has not all been driven by the membership. The local community health centre, for example, approached WTAA to organize an event to unite residents of different ethnic backgrounds around the enthusiasm demonstrated for decorating their homes for Halloween. The ensuing Great Pumpkin Art Festival with its professional, artist-led activities (e.g., pumpkin carving, mask-making, decoration-making, and relief printmaking) has now become an annual fixture in the community calendar. Another annual community celebration in the autumn is the Scarecrows in the Apple Orchard event hosted in Colonel Samuel Smith Park in New Toronto. This event began as a local collaboration between a puppeteer and a housing co-operative to draw attention to a historic apple orchard that was in danger of becoming a parking lot. Local historical knowledge and stories about food and the harvest were shared to help housing cooperative residents get to know one another, and scarecrows were built and paraded down to the apple orchard to be strapped to the trees as protective centurions. The WTAA has also helped to build social bridges between local historical societies and high school students in the Famous Historical People of the Lakeshore event, in which community workshop participants created large papier-mâché masks of historical notables from Etobicoke that continue to be worn in the annual Christmas parade. Together, the community arts events organized by the WTAA demonstrate an attentiveness to local potentialities and needs, as well as a desire to reinforce local optimism and to foster a deeper community awareness of local histories and geographies.

Franklin Horner Community Centre

North and west of the Waterfront Trail Artists Association, in the midst of manufacturing plants, warehouses, and working-class houses, is another

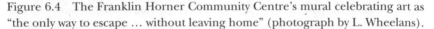

Figure 6.4   The Franklin Horner Community Centre's mural celebrating art as "the only way to escape ... without leaving home" (photograph by L. Wheelans).

unique community cultural hub, the Franklin Horner Community Centre (FHCC) (Figure 6.4). As a former 100-year-old elementary school that was converted into a community centre in the 1980s, the building has gradually evolved into a social, recreation, leisure, and sports facility for people of all ages. Its membership base of 1,200 is drawn from the Filipino, Polish, Italian, Serbian, Mexican, and Anglo communities. Without permanent city funding, this community centre relies on a $20 membership fee and the unpaid labour of many of these members, particularly older adults, to renovate and repair the building, and to help run fundraising events. A grant from the Ontario Trillium Foundation has helped to increase the public visibility of this community centre by supporting the construction of a colourful website that profiles the activities, events, and groups hosted here (www.franklinhorner.org).

The FHCC hosts 52 member groups. These include Lakesteps Narcotics Anonymous, Ontario Early Years daycare program, Colin's taxi-driving school, the Humber Barber Shop Quartet, the Polish Eagles dance troupe, and the Filipino Folklore society. Indian wedding and anniversary parties are held in the gym, as well as after-school tutoring and sports programs; Mexican dance troupes train in a dance studio with specialized sprung floors that is also shared by yoga students; a 17-piece band practises in the auditorium; seniors lunch-and-learn programs happen in the craftroom; the woodworking machine shop is busy all week; a former girls' bathroom has been converted into a darkroom for photography classes (and was featured in a British fine art photography magazine); fine arts studios are rented on a permanent basis by professional artists and used on a more temporary basis for photo shoots and life-drawing classes; and the local arts service organization, Lakeshore Arts, has moved in as a cost-saving initiative. The long-term goal of the manager of the FHCC is to make the centre an arts teaching facility. This vision demonstrates that the purpose of the centre is not merely passive art consumption but active engagement, so that a deeper connection to the creative process, from amateur to professional, can be fostered through participation in cultural production.

Culture Rising and the Independent Music Scene in Streetsville

The desire to augment community cultural participation is also in evidence beyond the western boundaries of Etobicoke, in Mississauga. Much as in Toronto, the idea that culture needs to be spatially incubated is asserted in Mississauga's cultural policy documents. Although the *Culture Master Plan* for Mississauga briefly specifies that "grassroots infrastructure is lacking" and that "incubator and generator spaces are required for cultural development and sustainability" (City of Mississauga Culture Division, 2009, 31), nowhere in the document are the terms "incubator" and "generator" defined or examples provided of their spatial expression. The plan does identify Streetsville as a potential "cultural node" and a place where "cultural resources can thrive," because the infrastructure database shows that over 50 cultural spaces are already located there (City of Mississauga Culture Division, 2009, 48). While this cultural policy assessment of Streetsville as a "cultural node" may be accurate, the over-reliance on numerical concentration data hides any intricacies of how the act of cultural incubation occurs on the ground in specific locations.

Streetsville is one of Mississauga's nineteenth-century "urban villages" (Emerson & Emerson, 2002). Today its heritage is proclaimed with street banners "celebrating 150 years" and faux-rustic wooden roadside signs that welcome visitors to "the village in the city" where "small town history is in every step you take." The historic village character distinguishes itself from the more conventional suburban built form with a grid street pattern of short, walkable blocks, a main street of stable and varied retail activity, and an interwoven mix of new and heritage buildings of various types and uses. A GO Transit commuter train/bus station provides an important public transportation node for a relatively dense residential community. In place of chain or big-box stores, the main street, Queen Street, hosts a variety of independent retailers including pubs, restaurants, cafés, bookstores, galleries, fine arts schools, and studios. Words like quaint, charming, and picturesque could easily be applied to this small town nestled within a vast suburb. What is less recognizable to an outside observer, and undocumented in Business Improvement Area directories or municipal cultural plans, is that Streetsville has a vibrant independent music scene.

At the heart of the contemporary music scene in Streetsville is Culture Rising, a community-oriented suburban arts project (Figures 6.5a and 6.5b). In its first incarnation it was squeezed between Fidalgo's Salon & Spa and Meena Cleaners, in a plaza with a bargain store, a franchise take-out pizza store, and a wine-making business. All that originally denoted this small indie art shop was a non-descript industrial glass and metal door leading down a narrow corridor to a cramped back room with no natural light. The store was opened in 2006 by two young band members to create a focal point for independent musicians, graphic artists, writers, and clothing designers. The owners "are committed to becoming a permanent stronghold for the growth, exposure, and promotion of local independent talent" in the 905 telephone area code region that surrounds the 416 area code of the city of Toronto (www .mississaugabongshop.com). In 2010, Culture Rising moved across the street from the plaza into a larger, more visible storefront location on Streetsville's main commercial artery, sharing sidewalk frontage with an adult video store and a salon. A second location was opened in Port Credit in 2012. The intent of both stores is to blend into the commercial fabric, not to generate noise complaints from neighbours or tenants, not to attract the attention of police, and to do what they do as openly and legally as possible.

Both stores make their money as head shops. They specialize in selling body jewellery and the drug paraphernalia used to consume marijuana. They sell bongs, vaporizers, roach clips, rolling papers, scales, cigarette lighters, and incense. In the words of one staff member, "glass pays the bills." But the store performs other functions too. Marketing postcards for concerts line the window ledge at the front door and tickets for

Figures 6.5a and 6.5b   Culture Rising's original storefront in a Streetsville strip mall and its new location across the parking lot on Queen Street South (photograph 6.5a by A. Bain and 6.5b by L. Wheelans).

local events can be bought at the cash register. A bulletin board at the entrance advertises guitar lessons, a band's need for a drummer, and an open mike night at a local pub. A magazine rack holds flyers for a needle exchange service, a local magazine (*Limelight*) promoting young artists in Mississauga, and independently published zines. A bin holds CDs of local bands for sale on consignment for $5. The walls are all lined with photographs, paintings, prints, purses, and T-shirts – artwork for sale by local emerging artists who are friends of the owners and staff. The store owners do not take a financial cut from the sale of any of the artwork.

Culture Rising is a self-described non-profit art gallery and "culture shop that sells and exposes items created by local artists to their primary audience" (www.mississaugabongshop.com). More than just a low-cost head shop that profiles local cultural products, Culture Rising is an informal networking venue that disseminates cultural information and connects creative practitioners of a particular age cohort and social circle with each other. The customers are regulars. Many of them grew up in the community and went to the local high schools; they cross paths at house parties and performances in home basements. Local emerging artists and projects have found one another and received a start here, neatly illustrating Watson et al.'s (2009, 860) claim that "creativity finds newness in both space and time through the mixing, encounters, and contacts between people and cultures."

Culture Rising became particularly important in 2007 after the only all-ages entertainment venue in Mississauga, the Masonic Lodge, closed because of complaints from local homeowners about vandalism and irresponsible behaviour. The Masonic Lodge had hosted hundreds of independent concerts and supported many emerging local bands. Four years later, a local promoter, the Big Scary Movement, was able to get the owners of the Masonic Lodge to host local music and live entertainment again with the understanding that there would be no loitering or littering (http://thebigscary.tumblr.com/). The only alternative to the Masonic Lodge is a local bar, the Duke of Marlborough, that is a popular stop for out-of-town bands and is known for heavy drinking and gritty, intimate rock 'n' roll shows. Pubs and clubs are important places for people to engage with live music, and they play a central role in the development of local music scenes (Watson et al., 2009). Across the city of Toronto "event promoters of all disciplines have complained about the lack of mid-sized venues" with the capacity to host 300 to 400 people (Morris, 2012). For under-drinking-age young people, there are one-off shows in home basements, churches, community centres, and youth drop-in centres every couple of months, but there are few places that host regular events or acts.

A municipality-wide music scene is nurtured by hosting consistent events in centralized and accessible venues. As one music promoter suggests: "If Mississauga had a venue in the heart of the city, right smack dab in the middle where it's accessible from all corners of the city, that would give a destination that could attract a regular crowd, which would, in turn, begin to develop a 'scene' again" (musician, Mississauga, 15 December 2007). An inadequate transit system and Mississauga's size tend to limit support for local music to particular small and fragmented pockets: Port Credit, an urban village along Mississauga's lakeshore, is known locally for its blues and country scene and Streetsville for its grunge and punk bands.

Since the mid-1990s, a handful of Streetsville bands have developed national and international reputations. The links between these bands and Mississauga are appreciated by resident musicians aspiring to follow in their footsteps, but the bands are not widely promoted by the city. The grunge band Billy Talent has been together more than 15 years, and, according to their website, it was the music of Nirvana, Pearl Jam, and Soundgarden that "first compelled four future chart-toppers from suburban Toronto to take up their craft in earnest" (www.billytalent .com). Notice that, in this quotation, Mississauga is not mentioned; it is subsumed into the broad swathe of suburbs that surround Toronto. Ill Scarlett is another band that formed as a hobby when three of the musicians were in high school. In a video interview with Rebecca Brayton (www.watchmojo.com/index.php?id=7532) she introduces them as a band "founded in Mississauga (it's near Toronto) in 2001" that plays what they call "pop-infused rock reggae." As the story goes, their name came from a themed Mississauga subdivision they passed when dropping off a recording at a bar. Between Erin Mills Parkway and Mississauga Road is Sherwood Forest Village, a subdivision with streets like Will Scarlett, named after characters in the story of Robin Hood. Not only are suburban houses and residential subdivisions with their juxtaposed social realities inspiration for song lyrics and band names, but the spacious basements of the homes are also often converted into rehearsal and studio space for cultural workers.

Home basements and storefronts like Culture Rising function as valuable informal cultural service hubs in Mississauga for a music scene that lacks an infrastructural backbone. Connecting these informal spaces are malleable collaborative networks, often nurtured in the virtual realm,

Figure 6.6   The Presentation House Theatre (photograph by A. Bain).

that help to contribute to the sense that the arts community is more than just a verbal construct. Creativity is fostered through unceremonious and unmarketed collaboration. As one musician (Mississauga, 15 December 2007) explains: "Most musicians within the community are approachable and support each other when they can. There is a sense of camaraderie coming from here. It's as if musicians band together in order to combat the problems." To create "from here" in a place that is often characterized as professionally and spatially isolated requires resilience and the support of an inclusive social network. Such unofficial networks underpin creative practice in cities and suburbs across the country.

The Creative Dominion Society in the Presentation House Arts Centre

In North Vancouver, the primary formal cultural hub is the Presentation House Arts Centre located just west of the main north–south downtown artery, on Chesterfield Avenue near 3rd Avenue (Figure 6.6). A former school, constructed in 1902, is the structural focus of the Arts Centre. The school was used as a temporary city hall as well as a police station before the North Vancouver Community Arts Council advocated for its conversion into a community arts centre with a gallery, museum, archive, and theatre. The North Vancouver Community Arts Council and the North Vancouver Archives have since relocated, but the North Vancouver Museum and Presentation House Gallery and Presentation House Theatre remain important cultural institutional tenants. An additional building on site is the Anne MacDonald Studio Building. Formerly an Anglican church, built in 1899, it later became a church hall and then a Scout hall, before it was donated to the city by the congregation in 1973 as a character space for the arts and moved to its current location (www.phtheatre.org/history.html). The studio building is a small, versatile space that is variously used for auditions, workshops, rehearsal, dance and yoga classes, as well as community events. The Creative Dominion Society (CDS) is one such regular user of the studio building for its signature bi-annual Launch Pad program.

The non-profit CDS was established in North Vancouver in 2003 by a local arts administrator who trained as an actor, dancer, and cultural critic and works out of his home office. He wanted to provide a community space within which to foster critical dialogue between artists and art audiences about interdisciplinary (especially performance) arts. Through a small operating grant from the Arts Office and in-kind sponsorship from Presentation House Theatre, the John Braithwaite Community Centre, and branches of the North Vancouver Public Library, CDS has temporary use of free space to host events.

CDS focuses on running a creative-process facilitation program, Launch Pad, for individual performers (e.g., actors, dancers, storytellers, masked performers, clowns, and writers) in different disciplines who create original works of art. The North Vancouver founder and artistic director of CDS and the Launch Pad program (6 June 2007) describes himself as a "creative consultant" and dramaturge who is interested in "how we structure work, how we create work, how we talk about it, and how we frame it when we actually present it to the community." He describes the CDS as "a kind of incubator or a vessel for artists." While many of these solo performers create much of their work in isolation, they often benefit creatively from the cross-pollination of ideas and the critical feedback they receive in developing their projects.

Twice a year, the CDS makes a call for artists to participate in the Launch Pad program via the email listservs of arts organizations in

North Vancouver and Vancouver. Once participants are selected by the five-member board of directors, the artistic director volunteers his time and subsidizes rehearsal space to work one-on-one with each performer for up to six weeks. As he explains: "They literally work in front of me and I watch and I witness, and either at the end of the studio time or by the end of the day, sometimes in a follow-up conversation or a coffee shop meeting or email, I offer them feedback ... we're always constantly thinking or talking about the work and we're trying to frame it in a bigger context" (6 June 2007). Once the front-end creative process is complete, flyers are then distributed to local coffee shops, community centres, and theatres advertising the pay-what-you-can event at the Anne MacDonald Studio Building. The rehearsal and performance space is donated by the Presentation House Theatre twice a year to host the Launch Pad program. No fee is charged for the artists to be in the studio or for the creative consultation process. Deliberately intended to be informal and accessible, the studio building affords CDS a relatively unintimidating, intimate, and affordable cultural space within which to engage a diverse community audience (of up to 40 people) in the creative process. Performers often bring their own friends, colleagues, and supporters to the event, and the performance is followed by a facilitated conversation in which audience members as well as performers are invited to ask, and answer, questions. Creativity, within the context of the work supported by the Creative Dominion Society, is celebrated as a human-propelled process of dialogue and cultural-knowledge exchange.

The Ballet de Printemps

Rehearsals and performances also drive the work of the Ballet de Printemps, but unlike the Creative Dominion Society, it has a permanent home. Established in 2007 by a former ballerina with the National Theatre of Croatia, Ballet de Printemps bills itself as Canada's newest professional repertory chamber ballet company and North Vancouver's only professional dance company (http://www.pentapix.com/live_ websites/ballet/home.html). It is located in a subleased converted warehouse building just east of the Lion's Gate Bridge and north of the railway tracks that cut through North Vancouver's industrial waterfront and shares a reception area with an engineering firm (Figure 6.7). The building's large size, open-space layout, affordability, and proximity to downtown Vancouver's dance scene appealed to its young, newly

Figure 6.7   Placard advertising Ballet de Printemps against the backdrop of a working port (photograph by A. Bain).

immigrated Croatian-Canadian founders. The sweat equity to retrofit this warehouse came from the artistic director and her technical director husband. Together they built changing rooms, a kitchen, and a lounge, all furnished with recycled office furniture, and constructed a sprung wooden dance floor lined with bars and mirrors that is now well used by the seven-member company and its 40-student ballet school. In the words of the artistic director (31 May 2007), the goal of the school is "to not only teach ballet but to give the young students, the next generation, the opportunity to learn about everything related to ballet as a theatre art including dance history, art history, music, acting for the stage and technical theatre." The school, together with the online sale of dance products, helps to fund an annual spring season for this professional ballet company. The dance company is deliberately small in order to retain sufficient mobility to travel with its roll-out floor and set to perform in communities throughout British Columbia.

Of the seven dancers on contract, three live on the North Shore and four commute from downtown Vancouver. Many dancers in North Vancouver describe the dance community on the North Shore as loose-knit, with most of the innovative and experimental work produced "in town" (multimedia performer, North Vancouver, 30 May 2007). Downtown Vancouver is perceived as the hub of the Vancouver metropolitan area's dance community. Ballet de Printemps' artistic director explains: "Professional dancers: they congregate downtown for a drop-in class or they rehearse in the dance centre all within one or two blocks. They're all very close. It's like the hub of everything that's happening" (artistic director, North Vancouver, 31 May 2007). It remains an ongoing challenge to create a dance hub in North Vancouver. None of the local theatres is equipped with a high-quality dance floor. Moreover, the five local dance schools and studios in North Vancouver are private and compete, rather than collaborate, with one another for students and sponsorship funding. Ballet de Printemps is unique as a suburban professional dance company and it functions as an informal hub for North Vancouver's growing ballet community. Attention now turns from North Vancouver to Surrey.

South Surrey and White Rock Art Society

As discussed in the previous chapter, the formal cultural focal point in the City of Surrey is the Surrey Arts Centre. There is, however, a limit to how well one cultural facility can serve a decentralized population. Among many grassroots, non-profit arts organizations in Surrey, two in particular have played a valuable role in nurturing the work of local creatives: the South Surrey and White Rock Art Society (SSWRAS) and Jarnail Arts.

The SSWRAS has had many name changes over its 50-year history, reflecting its geographic location and its lobbying strategies for political and financial support from the cities of Surrey and White Rock (www .artsociety.ca). Since it was established in 1959, it has grown from a handful of members to over 100, the majority residing in South Surrey. The membership comprises mainly amateurs and retired people with about 20 per cent being art professionals and younger students. For an annual membership fee of $35, members receive a monthly newsletter, participate in exhibitions, critiques, and the Summer School of the Arts, attend demonstrations, workshops, and courses by professional artists, and borrow art resources. Monthly meetings are advertised in the local newspaper and are announced by teachers in the art programs of public schools. These meetings are held in the general activity room of the local St John's Presbyterian Church hall, where large tables and a mirror on wheels are the only permanent fixtures. Other than a small library of art books, magazines, and DVDs that are lent to members, the society is unable to store any art supplies or props in the church hall. While exhibitions are occasionally held here, over the years the society has held exhibitions in the storefront of an abandoned bank building, a vacant group home for people with disabilities, and, more recently, the home studios of members, and the hallway of the Art Gallery of Surrey (where they are permitted one show a year). In the words of a long-time member of the society (artist, Surrey, 18 May 2007):

> We're here, there, and everywhere. We're kind of gypsies at this point, get-ting exhibition space where and when we can. We're meeting in a church and, because we don't apply for or receive any government grants, we've gotten to be very good at cost controls. We've become wonderful mooches. I mean, if we can find a way to do something for free or next to it, we pro-bably will, just because we have to.

This quotation captures the pride in the resiliency, thrift, and just-in-time solutions that this suburban community arts group has demonstrated in the face of long-standing spatial uncertainty.

For over a decade, the group has tried to get a permanent space of its own in White Rock. Although they came close to obtaining an aban-doned school building and an agreement for the structure to remain on the centrally located school grounds, a disagreement between the mayor's office and the local school board had the building bulldozed. Another time, they thought they might obtain space in a new high-rise development project. Frustrated with broken political promises in White Rock, and having raised more than $250,000, the society is now keen to collaborate with the City of Surrey to build an arts centre in South Surrey that would house an exhibition hall, classrooms, and studios. It has its sights set on the South Surrey Athletic Park. This recreation centre, with its many sports facilities, already offers art classes on a tem-porary basis in multi-purpose community centre rooms. The art society has worked with an architect to draft a conceptual sketch of an exten-sion to the existing community centre building. Such a permanent site, society members assert, would provide them with valuable visibility and a focal point. Although South Surrey is home to a range of arts groups (e.g., three life-drawing groups, a 70-member photography club, an 80-member quilters society, and a 50-member pottery group), specifically on the White Rock peninsula, there are also hundreds of independent artists who do not join clubs. The hope of society members is that "if there was a central focal point, maybe a lot of them would come out of their homes and studios and participate more, and share more" (artist, Surrey, 18 May 2007). Ever optimistic, imaginative, and unremunerated for their organizational efforts, members of the SSWRAS continue to scout adaptive reuse possibilities in suburbia.

Jarnail Arts

While the membership base of the South Surrey and White Rock Art Society is very Anglo-Canadian, Jarnail Arts is an informal cultural hub for the Indo-Canadian diaspora in the Lower Mainland (Figures 6.8a, 6.8b, and 6.8c). This multi-purpose arts school, gallery, and studio was established in 2004 by a Sikh visual artist and a dancer who emigrated to Surrey from a small town in the northern Indian state of Punjab. Jarnail Arts is located in a converted, two-storey, aluminum-sided office building that the artists bought in a small strip mall just west of the Surrey Arts Centre in central Surrey. Aside from constructing storage space on the ground floor to hold stage sets, costumes, musical instruments, and art supplies, very few modifications were made to the 140-square-metre building to accommodate its new cultural uses. The ground floor with its carpeted floors and high ceiling is used in the evenings and on weekends as an arts academy that offers Indian dance, music, and acting lessons. Past makeshift gallery walls lined with oil paintings depicting the traditions of village life in rural Punjab, and up a flight of stairs overlooking the performance space, is a narrow painting studio. Here on weekdays, the resident visual artist has created hundreds of paintings of key figures in Sikh history and scenes that reflect the struggles of the Gurus. Some of these paintings have been exhibited at the Surrey Art Gallery and at local public libraries; others adorn the walls of the Canadian Singh Sabha Gurdwara and homes in Surrey. A more public expression of this artist's work can be found in a large mural on the exterior wall of a seniors' housing unit nearby that documents the refusal by Canadian authorities to allow the ship *Komagata Maru* to dock in 1914. This mural serves as constant public reminder in Surrey of discriminatory and exclusionary Canadian immigration policies.

Jarnail Arts is a space of creation and education for a few professional practitioners and more than 75 Indo-Canadian pupils. Here,

Figures 6.8a, 6.8b, and 6.8c   Jarnail Arts: the entrance, dance floor/gallery, and painting studio (photographs by A. Bain).

cultural practices and knowledge are exchanged through lessons and rehearsals. Business cards, a website (www.jarnailarts.com, a small sandwich board at the entrance to the parking lot, occasional articles in ethnic newspapers, involvement on the board of the Surrey Arts Council, the design of city public banners, and the painting of murals, have all helped to make this Indo-Canadian cultural hub more visible to the wider Surrey community. Creativity in the context of Jarnail Arts is rooted in tradition and the transmittal of cultural forms from one generation to the next through community celebrations in a range of different formal and informal presentation venues.[5]

Valuing Informal Community Cultural Service Hubs and Community Arts

From the suburban cultural frontiers of Toronto and Vancouver this chapter has provided examples of a range of what I term "informal community cultural service hubs" that are variously embedded in the social life of Etobicoke, Mississauga, North Vancouver, and Surrey. A hub can be defined as a centre or focal point of activity. These examples demonstrate the importance in the suburbs of grassroots cultural workspaces that support the creative process, rather than the end cultural product, and that value the social rather than the economic potentials. These are spaces that have emerged organically rather than in a formally planned and funded fashion. They are centred on learning, the exchange of ideas, and constructive critical feedback. And they involve cultural workers who are employed in different creative disciplines and who are sometimes members of different ethnic communities. The intent of many of these non-profit initiatives is often to use community-based art forms to bring people together in a creative capacity. Cultural workers collaborate, often without pay, to maximize their creative potential, spatial control, and responsibility, and to build a sense of community.

One visual artist I interviewed talked about overcoming what she described as "my self-consciousness" (Mississauga, 30 October 2007) about admitting to others that she lives and works in the suburbs. She did so by publishing a community arts magazine that brought her into contact with other local arts practitioners. Her story is not uncommon. Of the 33 female suburban cultural workers in my research sample, regardless of career stage, 24 were directly involved in a visible and continuing way with the local arts community (e.g., through workshops, performances, exhibitions, events organization, committee work, or activism). Of my cultural worker study participants, 73 per cent of females and 40 per cent of males included local community involvement as a significant part of their artistic practice. For some, as practising professionals, such public service in grassroots community arts initiatives contributes a degree of legitimacy to their occupational identities.

Within my research sample, some cultural workers mentioned peers who deliberately distanced themselves from suburban community cultural events that might be stigmatized as amateurish and unprofessional. A printmaker from Etobicoke (25 October 2006) talked about the fine line that divides amateurs and professionals and the range of community involvement she has witnessed in the local arts scene:

> We have an artists' co-op just down the street and they keep to themselves. I have a friend who lives there who is a professional artist and has nothing to do with community art, nothing at all. He has lived there for seven years and he is a member of Open Studio and he does everything downtown. He doesn't want to be part of the local community because he considers it not to be a serious community, in the amateur sense. There's a perception by some of the professionals that the local community isn't worthwhile getting involved with. And yet there are other professionals who are involved. I'm in the middle because I think that local arts are important, the community arts, and the amateurs and semi-professionals. We're all part of a community that supports one another. They're being a little myopic saying they're not professionals. I think that is not good because those are the people who come to the performances and come to the art shows.

In my research sample, the distancing that this printmaker describes was most common among male cultural workers, particularly established professionals or those at mid-career. These individuals sought instead to reinforce their occupational identities by actively maintaining professional networks outside the suburbs; such networks comprised strictly urban, national, and international linkages. I suggest that it is the artists, writers, and performers who are more willing to challenge the boundaries between different categories of art – high/low, professional/amateur, and individual/collaborative – who perhaps have the greatest role to play in regenerating suburbia through informal place-making initiatives (Ford-Smith, 2001).

In an essay on the politics of community art practice, the Jamaican-Canadian playwright, actress, poet, and scholar Honor Ford-Smith (2001, 14) writes that in "regions where contexts are more grassroots and cultural infrastructure less textured, artists have closer ties to the local and to social issues around them." This quotation neatly describes the situation in many Canadian suburbs where cultural workers are often deeply embedded in communities, forming close working relationships with local residents that celebrate local histories and build local networks of support. Professional producers of culture, Carol Becker (2009, 65) asserts, are "often adept at trusting a personalized *experience* of place and then acting on it," using "their intuition about what might work as a creative response in a particular context." It is in suburbia that cultural workers perform the valuable function of filling in suburban interstices with small-scale projects that embed the arts in everyday routines and spaces.

Over 25 years ago, David Listokin and Patrick Beaton (1983, 170) noted that "the incidence of suburban cultural activities-facilities has grown," most especially in the older, mature suburbs. They suggested that arts activities tended to cluster in the older rather than the newer suburbs because of the available range of physical facilities that were closed or marginally used such as movie houses, churches, schools, and stores. The current shortage of single-use cultural facilities in the suburbs means that cultural workers and arts organizations often partner with alternative venues and develop innovative strategies to showcase their activities.

In the examples in this chapter, warehouses, storefronts, schools, churches, and recreation centres have all been variously transformed, used, and imagined as hubs of culture for diverse communities of practice that span the continuum from amateur to professional. Some cultural workers adaptively reuse spaces on a provisional basis, as temporary interventions in, and contributions to, the suburban cultural landscape. Others have created more permanent spaces.

The existence of these informal cultural spaces shows that the suburban built form is malleable and changeable. Each structure that suburban cultural workers have reconfigured with their sweat equity and ingenuity "has its own personal timeline, its wealth of forgotten histories and contexts, each use somehow informing the next while altering the building in its wake" (Christensen, 2008, 1). Suburban cultural workers are reinventing and rediscovering the spaces of suburbia, showing that buildings, like their urban counterparts, have lives that can extend beyond their original function (Montgomery, 2008). The spaces and layouts of buildings, as Kolb (2011, 163) emphasizes, can be reinterpreted in ways that open up "more possibilities for movement than the current social grammar dictates for its use. Dense city spaces encourage unforeseen encounters that breed innovation. Dense but spatially non-contiguous connections in the suburbs can do the same."

The process of arts-led redevelopment in the central city, instigated by artists and documented by scholars, is now gradually finding expression in suburbia. Not unlike what occurred in the deindustrializing centres of cities, when regulation decreases, interest in possibilities increases; this, in turn, renders different spaces attractive to different people for different uses.

Suburban cultural organizations and practitioners are particularly well-positioned to appreciate the various creative possibilities that the suburban landscape affords because of the long-standing cultural infrastructure inadequacies that they have contended with. However, a "lack of suitable spaces often leads to adoption of whatever space is available, which often brings with it limitations of function" (Duxbury, 2008, 33). A drawback of using buildings designed to serve multiple purposes is that not all purposes are necessarily well served. Invariably there are compromises in quality and functionality. Nevertheless, it appears that a new generation of cultural practitioners, with its "multidisciplinary cultural practices, preferences, needs, and expectations," is "less interested in large, formal, single-use facilities and more interested in multidisciplinary facilities and smaller arts spaces" (Duxbury, 2008, 9). The emergence of informal community cultural service hubs in the suburbs both indicates and speaks to a need for smaller, multi-purpose cultural spaces that encourage an intensification of activity and create opportunities for temporary usage (Bishop & Williams, 2012).

The value of the service hubs assessed in this chapter comes not just from the artful reuse of space, but also from the cultural workers themselves, who see the possibilities and seek to realize their visions for alternative suburban futures. As Doreen Massey (2005, 13) articulates, it is the "liveliness, the complexity and openness of the configuration itself, the positive multiplicity, which is important for an appreciation of the spatial." Thus this chapter demonstrates that, rather than looking for completeness and closure, many suburban cultural workers revel in the unpredictability of new uses and possibilities in suburbia and, in the process, may increase local awareness about the structural and lived complexity of suburban places.

Conclusion: Cultural Workers as Suburban Place-Makers

The question at the core of this chapter is: how do cultural workers contribute to suburban place complexity? Following Kolb's (2008) lead, the words *space*, *interpretation*, and *participation* can frame an answer. Socio-spatial arrangements are a significant component of the structural complexity of places. Through creative and spatial interventions, cultural workers have the capacity to challenge the linearity of places by reinterpreting the built environment to allow for different patterns of use and interaction. This redefinition of place norms and spatial structures – a reweaving of the urban fabric – is central to the act of place-making. While it is usually the professions of urban design, urban planning, and architecture that are most closely associated with the practice of place-making, it is now widely acknowledged that the arts augment a community's sense of place (Hudnut, 2003).[6] The place-making skills of cultural workers are invaluable in that they address misperceptions of suburbs as disfigured, disconnected, and monotonous. Through the arts, cultural workers can extend the horizons of spatial and social possibilities and meanings by reimagining alternative futures for places. By helping to fill in the cracks of the in-between city with small-scale projects that "integrate the arts with the more familiar, routine functions" of suburban life, cultural workers can contribute to a suburb's complexity (Hudnut, 2003, 196).

Place complexity is also fundamentally rooted in the generation of meaning through interpretation. As Ronald Fleming (2008, 14) details in his book *The Art of Placemaking*, "Place is not merely what was there, but also the interaction of what is there and what happened there ... it is the recollection of patterns of life lived in a particular building or space that creates the 'cornerstones' of mental association and gives such places the patina of affection." During the creative process many cultural workers seek inspiration from their local surroundings and the art they produce may, in turn, contribute to the "historical density" of suburban places by revealing and inscribing traces of narratives and memories sheltered therein (Kolb, 2008, 73).[7] In this way, cultural workers become vernacular memory-makers who help to build "community cohesiveness" by shaping the past into a meaningful present (Rodriguez, 1999, 2). As I have argued elsewhere (Bain, 2006), cultural workers (particularly visual artists) are a social group with very strong emotional and physical attachments to places and they have the potential to challenge the material and imaginative disappearance of places at the neighbourhood scale. In suburbia, cultural workers, through imaginative, interpretative, and spatializing acts, help to create points of informal social interaction that link places and foster greater community awareness among local residents.

Chapter Seven

Networked Geographies of Suburban Cultural Workers

Instead of a geographical art scene, I have Twitter buddies and Google Reader.
—(Kleon, 2012, 89)

The dominant assumption in cultural planning and policy circles is that economically successful and innovative "cultural production is based upon the development of complex and extensive – and hence largely urban – agglomerations in the form of cultural districts" (Bader & Scharenberg 2010, 78). Clustering is reasoned to be a key ingredient in the economic viability of cultural districts and creative industries. Consistent with the scholarly literature on industrial districts and agglomerations, clustering is said to support networking and the exchange of information, knowledge, and expertise. Montgomery (2008, 73) explains that a "key feature of the creative industries is that they are not only 'hi-tech' but also 'hi-touch.' This means that personal lifestyle, the ability to mix work and play, the opportunities for personal contacts and networking in less formal surroundings – bars, cafés, restaurants, clubs – are of paramount importance for the creative industries." However, part of the challenge of living and working in suburbia is that there are few physical networking sites, and those that do exist are usually spatially dispersed and established by non-profit organizations with unusually diverse memberships. The notion that clustering occurs because of shared customers, suppliers, labour, and technologies is rarely evidenced in the suburbs. As discussed in chapter 5, a cluster-based cultural economic development model has little utility in suburbia. It limits and distracts suburban decision makers from the important project of developing the conditions necessary to sustain the everyday creative practices of local cultural workers.

The phrase "splintering urbanism" energetically conveys the forces of fragmentation at work in contemporary cities and alludes to the need for mechanisms to assist city-regions to retain their cohesiveness (Graham & Marvin, 2001). This chapter highlights the value of a social network–based cultural economic development model in the face of spatial fragmentation. "Most critiques of suburbia share the presupposition that spatial proximity remains our most important mode of connection ... But spatial proximity is becoming just one mode of connection and community affiliation; there are more modes of contact available, and these can bring complex encounters and self-redefining innovations" (Kolb, 2008, 148). This chapter argues that the networks formed through computer-facilitated communication help to sustain the dispersed geographies of cultural production and cultural consumption in suburbia. While the virtual cultural realm may not garner the same arts world respect as its arts district counterpart, this chapter argues that the computer and the internet are important tools of interactive communication and connectivity for suburban cultural workers, providing valuable online communities of support and collective learning (Castells, 2000).

The Social Morphology of Suburbia:
A Shortage of Spaces of Encounter

A common perspective on cities is that they are places of encounter and arenas of social interaction (Amin et al., 2000; Simosen, 2008). Lefebvre (1991, 101) has theorized that "urban space gathers crowds, products in the markets, acts and symbols. It concentrates all these, it accumulates them. To say urban space is to say centre and centrality." While cities may be characterized as places of encounter, interaction, accumulation, and centrality, that same characterization is seldom applied to the suburbs. The suburbs are more commonly depicted as spaces of privacy, predictability, and dispersal than as spaces of unanticipated encounters and centrality (Sibley, 2001). As discussed in chapter 4, the theme of privacy was prevalent among the advantages cited by suburban cultural workers for living and working in suburbia.

The sociologist M.P. Baumgartner (1988) determined from his one-year study of conflict in an American suburb that suburbs are physically and socially configured in ways that permit significant privacy and separation. Baumgartner (1988, 3) discovered that "it is not uncommon for people to know few of their fellow residents. More important, suburban households often are separated by a great deal of social distance." In light of the high transiency rates he uncovered, Baumgartner (1988, 9) characterizes suburbia "as a place where comparatively 'weak ties' to other people, rather than strong bonds of enduring and great attachment, typify social relationships." He concludes that suburbia has a distinctive social system that "combines transiency with homogeneity, and autonomy and independence with a relative absence of strangers. It entails a diffuse kind of interpersonal association, marked by considerable fluidity and distance in relationships" (Baumgartner, 1988, 10). A desire to limit interactions with neighbours and to protect home workspaces from unwanted intrusions from family and friends has been expressed to some degree by the majority of cultural workers in this study. "Transient," however, is not a useful descriptor of the suburban residencies of cultural workers. In this study, many suburban cultural workers have spent much of their adult lives living and working in suburbia. There are those in the research sample who have grown up in the suburbs, attended satellite university and college campuses, and have then settled there after graduation. There are many more who have moved to the suburbs following their post-secondary education to become homeowners and/or to start families of their own. The private home is the centrepiece of many suburban cultural workers' social lives.

Suburban street life is often undeveloped because people choose to travel between home, work, errands, and social outings within the enclosed confines of private transportation. The American urban sociologist Ray Oldenburg has persuasively argued in his book *The Great Good Place*: "What suburbia cries for are the means for people to gather easily, inexpensively, regularly, and pleasurably – a 'place on the corner,' real life alternatives to television, easy escapes from the cabin fever of marriage and family life that do not necessitate getting into an automobile" (http://www.pps.org/reference/roldenburg/). As the discussions of cultural infrastructure inadequacies in Chapters 4 and 5 illustrate, there are rarely any identifiable or officially designated arts districts in the suburbs. Furthermore, there are few places where people can meet regularly to engage in critical cultural dialogue. One long-time resident and arts educator in Surrey (3 June 2007) bemoans:

> There's no arts centre or arts district. You can't go out and have a meal and
> go to the theatre as you do in Vancouver. You can't go to a performance
> and then go out and have a cup of coffee or a late-night snack afterwards
> to think and to talk to other people about it. You go, you listen to a performance, and you go home. There's no sense of pulling people together as

a whole. The scattered people come together, enjoy the performance, and then go home or maybe one or two will go to someone else's home.

The cultural event, be it in a suburban arts centre, theatre, performance hall, banquet hall, community centre, library, school, or church, becomes the moment of encounter between cultural worker and audience. Apart from attendance at the event and the privacy of home-based socializing before or after it, there are usually few opportunities, especially in the outer suburbs, for more informal public encounters in "third places" (Oldenburg, 1991).

An arts educator and a practising professional painter in Surrey (3 June 2007) explains:

> By nature, visual artists anyway tend to be individual. They tend to work on their own. They tend to get together when they have shows. The art gallery tried, I thought it was quite innovative, to organize what they call an Artists' Café which is a monthly drop-in time. You come in and talk about art a bit. It worked for a while. But what artists really need is a bar like the bar in New York City where all the abstract expressionists hung out.[1] You'd have a drink and you'd go back to your studio. You would take somebody else back and you discuss the work. It just doesn't happen here because there's no local focus. The Artists' Café didn't work simply because it was too far out of the way. You had to say "I'm going to spend the night going there, taking time off my studio," rather than saying "I'm taking a break from my painting, I'm going down there to have a beer, talk to the other guys who might be there, argue a bit, maybe go back and do some more painting." It's a whole different dynamic. It might be different in other communities which are smaller … There's no sort of ongoing, loose-knit communication as you would have in Vancouver where you walk along the street to a café, which is near your studio or your house, and you run into somebody else who is coming out of his place. It's very different. I'm not sure what the answer is, other than developing strong centres in all of the various areas of Surrey because you don't have any other way of really connecting to those communities. You can't always drag people downtown.

Social spaces – little gathering points – are important additions to the fabric of cities because, for many cultural workers, aside from their work-space, there are few other places to informally gather. More importantly, such spaces also help to give physical expression to the arts scene, demonstrating that the local arts community has substance and is more than a verbal construct.

Ray Oldenburg (1991) popularized the informal neighbourhood gathering places that anchor community life outside of home and work (e.g., café, coffee shop, pub, post office, main street) with the term "third place." He romanticizes these social spaces as the hubs of grass-roots democracy. Third places "host the regular, voluntary, informal, and happily anticipated gatherings of individuals beyond the realms of home and work"; they are places where conversation and discussion happen and where the rudiments of policy begin to take shape (www.pps.org/reference/roldenburg/). In the tradition of Paris's Montmartre, New York City's Greenwich Village, and San Francisco's North Beach, Richard Lloyd (2006) writes about the crucial role performed by bars, restaurants, and coffee shops in the emergence of an art scene in Chicago's Wicker Park neighbourhood. Lloyd (2006, 104 and 109) asserts that "new spaces in the neighbourhood heightened the visibility of local creative efforts" and "allowed for interactions that fostered a sense of community, and opportunities for mutual support and collaboration among aspiring cultural producers." His observations about the significance of social gathering points to the nurturance of creative processes in Wicker Park parallel the value of community cultural service hubs to suburban cultural workers discussed in chapter 6. Certainly, the suburbs, with their more restricted arts audiences and more limited cultural resources, could benefit from a "social infrastructure" that provides places to play as well as places to work (Hume, 1997).

The inner suburbs of Etobicoke and North Vancouver, more than the outer suburbs of Mississauga and Surrey, have a greater range of third places. Locally owned bars, coffee shops, restaurants, galleries, artist-run centres, dance studios, art supply stores, music stores, and book stores dot the suburban landscape, playing a valuable role in sustaining local cultural social networks. Many of these independent businesses, in addition to providing community notice boards, also rotate exhibits of local art, showcase local performers, advertise and sponsor local cultural events, and employ local cultural workers. Moreover, they contribute to the cultural economy by promoting and nurturing collaborations among creatives who informally interact with each other as colleagues, mentors, and gatekeepers within these spaces. Although such third places are variously known, appreciated, and used by local professional cultural workers and amateurs, they are spatially dispersed and do not exist in the same concentration as in the downtown core of large cities. With a shortage of physical spaces of encounter in suburbia, the virtual realm, and the networking opportunities it affords, plays a useful role in suburban cultural production.

Networking the Cultural Economy

The American sociologist Howard Becker (1982) in his book *Art Worlds* argues persuasively that art is a product of collective action that involves the labour, coordination, and creativity of many individuals. Cultural production at the scale of the individual practitioner rarely occurs in a neat commodity chain. The labour practices of individual cultural workers are perhaps better characterized as "circuits of culture" in which the movement of commodities through production, distribution, and consumption phases is envisaged as a non-linear circuit (Leslie & Reimer, 1999). The circuits approach "refuses to recognise beginning and end points in this movement and disrupts the priortisation of one moment of commodity circulation over another" (Hughes, 2000, 177). However, the network is perhaps the most appropriate metaphor for understanding cultural production. Over the last decade, the spatial metaphor of the network has been variously applied to the post-industrial cultural economy in order to appreciate the complex ways in which different kinds of nodes (e.g., people, organizations, businesses, and regions) are connected to one another in the production of culture.

Networks are an established topic of study in the disciplines of sociology and communications, where social network analysis is often employed to examine patterns of connection and exchange and the flow of information and resources among people and organizations (Wasserman & Faust, 1994; Emirbayer & Goodwin, 1994). Social network studies have detailed how people can be variously positioned centrally or more peripherally in interrelated constellations of connections that differ in size, heterogeneity, and purpose (Strangleman, 2001). The connections between actors in networks "are seen as complex 'webs of interdependence' rather than fixed, vertical, and uni-directional relationships" (Hughes, 2000, 178). According to social network theory, networks are localized spatially and temporally; they work in real places at specific times (Hughes, 2000). Although information and communication technologies (ICT) have lengthened networks, permitting them to span vast physical distances, it is important to remember that networks are rooted in physical places through material flows and social relations (Park, 2004).

Ideas about networks have been advanced by the urban sociologist Manuel Castells (1996), who first proposed the concept of the network society in *The Information Age: Economy, Society, and Culture*. In this influential book, Castells argues that in the last two decades of the twentieth century globalization and innovations in microelectronics-based information and communication technologies have radically transformed the structure of societies and economies around the world. The concept of the network is used to convey the way in which social relations have been stretched across space, producing flatter and more dynamic forms of economic and social organization. Castells (2000) explains that while networks are a traditional form of social organization, they are now increasingly complex and empowered by new communication technologies to cope better with "flexible decentralization" and shared decision making. Networks have the ability not only to dissolve centres but also to "disorganize hierarchy, and [to] make materially impossible the exercise of hierarchical power" (Castells 2000, 19). When information circulates in a virtual network structure, it can flow with little distinction or impediment between social, cultural, economic, political, and physical centres and peripheries. There is power in the movement of information and knowledge, and people are empowered through their access to them.

Inspired by the work of Castells (1996; 1997; 1998) and other sociologists, scholars of the cultural economy have begun to illustrate the essential role performed by elaborate networks of business and personal contacts in cultural production in large urban centres (e.g., Norcliffe & Rendace, 2003; Vinodrai, 2006; Stolarick & Florida, 2006; van Heur, 2009 and 2010). Such networks, however, can be a challenge to study because, as van Heur (2009, 1533) notes, "nodes of these networks are informal, temporally limited and partly hidden." Nevertheless, research on cultural production networks has shown that, although these networks are governed by internal logics and goals, they are often variously used to (re)confirm business and social relations, identify employment opportunities, establish new contacts, facilitate cooperation, obtain information, access resources, produce knowledge, circulate ideas and policies, and transcend local limitations (Coe, 2000). Much social science research and many government-sponsored reports focus on formal cultural networks that produce value-creation chains with quantifiable economic outputs. However, the informal networks of creativity and the people who actualize them also need to be acknowledged as important anchors in local suburban arts scenes (Bader & Scharenberg, 2010). Richard Smith (2003, 32) reminds us that "it is people who make up networks, people who make the ideas in networks, and people who put those ideas into practice." People create and maintain networks, giving them their dynamism and character. Inevitably, the shape and focus of networks shift in response to the people and the groups who are involved in them and the social relationships that evolve from them (Cumbers et al., 2008).

It is not enough, Coe (2000) argues, to recognize the presence of networks; it is also important to appreciate different types of networks and their qualities. The economic geographer Gernot Grabher (2006) has developed various classifications of networks. In one classification, he has characterized networks based on their duration and governance. Within the category of long-term networks he distinguishes between hierarchical and heterarchical networks. Hierarchical networks are controlled by a coordinator who regulates network rules, practices, membership, and resource allocation. On the other hand, heterarchical networks regulate interaction between members in a more distributed and associative way, exhibiting "patterns of emergent self-organization, strong lateral ties, and a diverse distribution of authority" (Grabher, 2006, 168). Suburban cultural workers variously participate in different networks that could be classified as either hierarchical or heterarchical.[2] The informal community cultural service hubs described in chapter 6 succeed in large part because of their heterarchical organization.[3]

Suburban cultural workers in this study articulate an attraction to networks where communication channels and information flows are reliable, open, and permeable. Some networks are more formal in interaction and hierarchical in structure, managed through websites and listservs run by professional and non-profit arts organizations, and formed through shared ties of experience and expectations. Other networks are more informal and social and heterarchical in their organization; individuals may offer to obtain and interpret information for each other as well as to mobilize collective resources. Many of these networks within the cultural sector may overlap, with their various members interacting and exchanging information and resources.

The social networks in which suburban cultural workers participate are differentially placed and resourced and vary in their geographic reach. In general, individuals who have immigrated to Canada[4] or who may be more established in their professional careers tend to have more national and international connections to colleagues, while younger, emerging artists may participate in more localized occupational networks. Regardless of whether an individual's professional relationships are local, regional, national, or international, cultural workers with fewer professional network linkages are more likely to feel marginalized within a suburban locale. Networks help cultural workers to connect with other like-minded cultural producers and audiences in a professional capacity while retaining their creative autonomy. The remainder of this chapter focuses on the role of computer-facilitated communication in the formation of cultural production networks in suburbia.

Spatially Dispersed, Yet Connected: The Digital Information and Communication Revolution

In the early 1990s, the World Wide Web was launched. A collection of documents and resources that are connected by hyperlinks and URLs,[5] the World Wide Web is an application that runs on the internet, a global system of interconnected computer networks. The World Wide Web and the internet, together with the hardware and software to store, transmit,[6] and network digital information, have formed the world's largest computer network (Mitchell, 2000). Physically created from computers and their networks, networking hardware, software applications, coaxial and fibre-optic cables, wireless communication systems, and communications satellites, this worldwide digital information infrastructure has rapidly reconstituted the relationship between people, information, and places.

In the twenty-first century, the effects of computer-facilitated communication systems have been powerful and widely felt. Computer-facilitated communication is the process through which people create, exchange, and perceive information through the use of networked telecommunications systems (Pickerill, 2003). The term refers to a range of communication formats that occur through computer-based technologies, such as the internet, email, listservs, chatrooms, blogs, and text messaging. This form of communication has transformed urban space, work practices, and everyday social interactions, creating new linkages between people and organizations within cities and between cities. The digital information and communication revolution has reconfigured city-regions, making centres out of peripheries. It has helped to make the suburbs a viable and attractive location of work and residence for "footloose" cultural workers (Frey, 2004).

Spatial proximity is not the only kind of configuration that can support innovation and creativity in the cultural sector. In suburbia, where cultural production is dispersed, creativity, innovation, and information exchange are fostered, in part, through computer-facilitated communications networks. The internet, websites, and email are all useful networking tools for suburban cultural workers; they are technological vehicles through which cultural knowledge is disseminated and connections made.

Although there was some variation in informants' online abilities and activities, all of the cultural workers and arts administrators interviewed for this study recognized the importance of being able to manoeuvre in the virtual social and cultural landscape of the internet. Rarely viewed with scepticism or suspicion, the internet is almost uniformly regarded as indispensable to creative practice and cultural administration. The cost to purchase a computer and internet service provision was rarely cited by suburban cultural workers as a deterrent to online participation.[7] It is more likely that lack of technical proficiency would impede the realization of particular online projects (although computer literacy and competency skills were quite high among cultural workers interviewed). Where more advanced technological competencies are needed (e.g., to write and design web pages that are engaging, dynamic, and visually pleasing), cultural workers often outsource this task.

The Internet

The internet and digital technologies allow cultural workers who can afford the cost of computer equipment and service provision to maintain social and economic relations that span the globe. Through interactive computer-facilitated communication cultural workers develop virtual connections with people, organizations, and markets located in places far from where they physically live and work. The internet is an accommodating virtual space that fosters "self-directed networking" (Juris, 2005). It shrinks the perceived distances between places and permits the free association of people and ideas in different combinations and in different networks.

For many suburban cultural workers and arts administrators the internet functions as a useful research tool. In the process of selecting plays to perform at the Presentation House Theatre in North Vancouver, the manager/producer (25 May 2007) "check[s] out websites of other theatre companies in Canada to see what their theatres are and just make a note of the sites. I look for the plays and see if I can find them and read them." Her online research of other English-language theatre companies, the plays that they have performed, and the reviews that they have received, informs the diverse programming at this intimate theatre venue. Thus, internet research can streamline the cultural production planning process. In Mississauga, a private gallery owner wanted to initiate an arts collective in her community. She approached this task by looking on the internet for other examples of art collectives and how they managed themselves. As this Mississauga gallery owner (5 October 2007) explains: "I went online and took a constitution from another organization and then I adjusted it for a Streetsville Arts and Crafts Collective." While imitation through virtual Xeroxing is perhaps not the highest form of innovation, these examples show the speed with which cultural knowledge can be appropriated and altered to meet local needs.

The internet has helped to level the cultural playing field and to increase the social porosity of suburbia. It gives individuals and arts organizations the opportunity to source information, resources, audiences, and support from a wide area. In the words of a curator and arts journalist in Mississauga (30 October 2007):

> The internet is a really great democratizer of media. People in Mississauga can make their own contribution there. When I was publishing a blog online after the magazine had stopped publishing, that is actually how I found people. One of the first few people that came on board to help me write was somebody whom I found when I searched Mississauga on YouTube and I found this amazing BMX video. He was one of the guys on the bikes. I emailed him asking him if I could write about his video, and he ended up being really interested in contributing. That has been a great tool. How would I have found these people otherwise?

Through computer-facilitated communication like-minded people, including those who may live and work within the same locality yet have never met, can connect. A photographer in Mississauga (11 November 2007) emphasizes how

> the internet makes it easier to connect with people. I was contacted early this year by a writer from the *Toronto Star* who wrote an article about that. She noticed that there was more happening in the suburbs basically because of the internet. It's very easy to connect with people and then of course work out of a home office. It's a lot easier to find information. The internet has been a great tool.

This founder and curator of the *A Month in Photography* festival reiterates the importance of the internet as a valuable tool for research and connectivity. In a *Toronto Star* article in which this curator is cited, the journalist Erin Kobayashi (2007) writes with particular reference to the Brampton Indie Arts Festival:

> Perhaps the greatest breakthrough in the last 10 years has been festival organizers' reliance on the Internet to network with artists and connect with audiences. The suburbs, with a reputation for being sprawling spaces full of single-family homes and a car-heavy landscape, are also seen as

intensely private. The Web allows those situated outside of the artistic hub of Toronto to inexpensively reach out to neighbours.

It is the internet that helps to socially facilitate many of the cultural happenings and arts projects in suburbia.

The social dimensions of connectivity are central to the process of creation. It is through a rhythmic alteration between solitude and interaction, between workspace and social space, that creativity is sustained (Buttimer 1983). Cultural production is supported not just by individual inspiration but also by the emotional and communicative support and stimulation provided by community. Using Howard Becker's work on the sociology of art as his starting point, Farrell (2001) shows the importance of collaborative circles and literary friendships to the development of literary careers, while in the cultural economies literature Stolarick and Florida (2006) demonstrate that increased interaction with peers can contribute to innovation and creative growth. The virtual world is crucial to the creative process because it helps some cultural workers to break out of their social isolation in a controlled and managed way. On the Canadian website Bookninja (www.bookninja.com) the author Brenda Schmidt (2005) writes about "causing" her own literary scene online:

> The Internet has opened the door for me as a writer and has left me standing in the doorway, neither inside nor outside of the scene. I never imagined I'd have the pleasure of sharing ideas with so many writers. As I see it, the Internet has placed isolation and the scene on opposite ends of a teeter-totter. Now, with the weight of each being more even, neither one is grounded, neither one is left swinging helplessly in the air.

But to access virtual arts scenes requires a grounding in a physical locale, as one North Vancouver writer (30 May 2007) suggests:

> The fact is that writers live inside their own brain and whatever you have that feeds that. You have to have a grounding in the real world. I firmly believe that. You have to be able to fill up the well with real-life experience because otherwise your work is shallow and uninformed. Because we're writers, that allows us to take that step into that acceptance of this ephemeral online existence. It makes it easier.

Physical existence provides this writer with the fodder for her writing projects and a stable platform upon which to virtually network with others. The lived realities of making a living and supporting a family all ground cultural workers like this writer in the suburbs, while connecting to distant others on the internet helps to release them from suburbia's potential isolation to imagine other possibilities (Cumbers et al., 2008).

Personal Websites

To become visible and to create a presence on the internet, many cultural workers interviewed, particularly visual artists, have personal websites. Creation of these personal websites is frequently outsourced to acquaintances with the requisite technical software and programming skills. The websites are then updated by the individual cultural worker. A painter in Surrey (24 May 2007) describes how she acquired and uses her website and how her computer literacy has gradually increased:

> When I first moved in [to the studio], I was totally computer illiterate. I really didn't want to get involved, but because [my partner] is, it looked interesting. He got me involved with the website and had a friend build it for me. Since then I've been learning how to use the computer and actually I quite enjoy it now. I keep trying to update my website with new work. But quite frankly between [my partner's] and my website, we really haven't had any traffic for buyers. I've had a lot of hits. We record our hits, people browsing and looking. But as far as anybody emailing me or actually asking me to buy something off my site, I've had nothing in four years. Which really amazes me because my agent has a professional art consultant website and she sells stuff off the Web. She's sold two or three pictures of mine off her site. The only people that look at my site are the people that get my cards or know me. I don't know what one can do to make it any better or different. [My partner] is the same way. He has photographs for sale off his site. He has PayPal and everything. Nothing. It's really weird.

Like this visual artist, many of the cultural workers interviewed who have personal websites initially thought that owning a website would be a way to reach a wider audience and to augment the sale of their work. Experience has taught them otherwise. In the words of an Etobicoke printmaker (25 October 2006), "I have a website and it does say I do custom printing on the website, but I really haven't gotten one order through it. It's more word-of-mouth." Another visual artist in Surrey (12 June 2007) echoes this sentiment: "I don't get much from the internet, but I do get queries." Most of his clients come from verbal recommendations rather than from his website. While sales for visual artists may occasionally occur via personal websites, for the most part their websites do not generate enough unsolicited audience traffic to garner new commissions. Sales for many of the visual artists in the research sample come through traditional word-of-mouth contacts rather than virtual networks.

In an effort to gauge the level of interest in their work, some cultural workers track the number of hits on their website that they receive. Others dismiss such counts as inaccurate and seek instead to record site visits instead:

> As far as hits are concerned, that is not a great benchmark to measure the amount of traffic that you have on your site because when the Google or Yahoo search engines find you, someone might be searching for kitchen sinks. I don't measure hits. I measure specific site visits. (sculptor, North Vancouver, 11 May 2007)

Despite the concern expressed by cultural workers that their personal websites do not generate significant business revenue or new audience traffic, they continue to use them and to pay the web-hosting fees.

Although personal websites may function ineffectually for sales, they do function effectively as a virtual storage space for work. Personal websites are used most commonly by research informants in an archival capacity. A photographer from North Vancouver (22 May 2007) explains how he relies on his website as "an online catalogue" of his cultural products that he can market to potential clients:

> I find it incredibly convenient to scan the negative and then just put up a low-resolution representation of that image on the site. It really gives the photographer a lot more scope to get his work out there, at least in a limited form, so that then people can see what's there and have some idea whether this is something that would be appropriate for their needs ... I rarely get feedback from the website. I rarely get orders from the website. But what I do with the website is that I use it as an online catalogue. My goal with it is not necessarily to market my work directly. My goal with it is to use it as a marketing tool when I am approaching different groups that I market to.

Websites are widely used by cultural workers to visually and textually display in digital form the range of work that they have produced. While the websites may not themselves generate new customers, they can be useful for displaying a portfolio of work to targeted clients as part of a broader marketing strategy.

Many personal websites can be quite basic in design and function. As one writer in North Vancouver (12 June 2007) admits, "It's a bad website. I don't update it often enough and it's not very dynamic. But I got it because people would write to me and ask me for something so simple, like a list of my books. Now I can do it on the website." For this writer, her website is an alternative-format curriculum vitae that she depends on for elementary information storage and distribution to interested people. The information that she can display and update on her website for people to independently access helps to reduce her correspondence workload. For an established painter in Surrey (3 July 2007), his website is also a simple access point for people interested in his work:

> It's easy for people to access my work. If they've seen it somewhere and they just want to Google my name, it comes up very readily. It shows galleries and my site. It's very easy to find. My website is a work in progress. I've been thinking about making it into more of a blog, probably with more information about current things. Mostly right now it's an archive. It's information in an archive.

The quotations in this section emphasize how cultural workers rely on websites as an affordable way to visually and textually archive their work. As small-scale electronic archives, websites are widely used to maintain network relationships by keeping former and potential clients and interested audience members informed of current creative projects. Certainly, with a shortage or absence of suburban art galleries, such websites can be an important venue for displaying and viewing works of art.

Another visual artist established her personal website in 2000 after building a mailing list of over 200 people from annual open-studio events, commissions, and print-media advertising. Even though she has had a website for a decade, this Surrey painter (24 May 2007) asserts that "the internet is still in the baby stages for me. I get the odd cold-call. I had a portrait commission from a lady in Victoria … The internet for me is the twenty-first-century calling card. I meet people, I give them my card, and if they want to see more, they just have a look at my website." Cultural workers feel pressure to keep up with their peers with respect to their technological knowledge. A Surrey writer (24 May 2007) describes the importance of remaining technologically *au fait* in order to maximize his audience outreach potential:

> A minute on TV will gain you more exposure and more accolades than your radio program. Radio will benefit you more than a newspaper story.

When you think about it, how do people acquire their information? The best of all possible worlds for me is to be on www.YouTube.com. If I was working harder at my publicity, I'd get myself a four-minute video on YouTube. It's amazing. It's so powerful. Kids now, they're tuning out TV and they're getting their information online. If I wanted to reach that audience, hands down, that's the way. Do something on the computer. Do something on the internet.

The computer has become the communication centrepiece in the twenty-first century. It supports the use of the internet for research and networking, participation in chat rooms, the management of a blog, a website, or a MySpace[8] page, and email communication. But for many cultural workers, particularly those over the age of 50, computing knowledge generally remains at an intermediate level at best.

Email and Membership Listservs

After the telephone, email is the most commonly used means of communication among cultural workers and arts administrators. For those cultural workers who may have a more solitary creative practice, the telephone and email are described as versatile and affordable ways to keep in touch with colleagues. Take, for example, the socializing experiences of a sculptor in North Vancouver (11 May 2007):

One thinks in terms of old Paris, the cafés and the bistros where the painters used to meet and talk and get drunk and visit. But a stone sculptor, especially, is a very solitary vocation or artistic process. You can't stick in a room as painters can and put easels back to back and visit and paint at the same time. You just can't do that. As far as meeting with others, I'll probably talk on the phone or exchange emails. But each of the associations I belong to have annual symposiums which are week-long events where we all get together and carve and share ideas. We usually rent a big soccer field so that we can bring a compressor and generate some design. On a daily or weekly basis, I don't really see those colleagues that much. I tend to gravitate towards friendships with the people who are more interested in talking about and collecting art than with other people who create art. We do have friends who are artists in other disciplines that we see now and then.

This stone sculptor describes the labour of his professional practice as a solitary endeavour. The opportunities for him to socialize with other artists are few. The telephone and email allow him to informally maintain weak ties with colleagues – ties that are later strengthened through face-to-face time spent working and interacting at annual symposia. Another cultural worker from North Vancouver, although she works in the different discipline of theatre arts, emphasizes the social value of email. After a decade as a professional actor (21 May 2007), running into colleagues at auditions and performances, "it's keeping in touch by phone and email and going to see shows in Vancouver" that allow her to stay connected to the theatre scene.

In North Vancouver's dance community, one ballerina and ballet company owner (31 May 2007) who immigrated from Europe retains her European professional ties through reading dance journals and corresponding with colleagues on the internet and MSN. "We chat about what's going on, performances." Interpersonal relations, Jenny Pickerill (2003, 110) has determined, "can be difficult via email, and online involvement can be transitory. Personal interaction online is often based on 'thin' understandings of identity, and commitment to online transactions can prove to be limited." Within the context of suburban cultural production, email can be useful for weak connections. Deeper, more meaningful interactions among suburban cultural workers may be more likely to take place at professional conferences, meetings, or other cultural events that afford opportunities for face-to-face engagement and conversations about shared experiences.

Email is perceived as a safe and undemanding way to maintain contact with colleagues and to initiate contact with new people. This computer-facilitated written communication is spam-filtered and self-regulated. Individuals can choose to ignore or respond to the email at their leisure. A storyteller in North Vancouver (21 May 2007) recounts how a colleague across the country electronically introduced herself via email:

> There's a storyteller in Toronto and she emailed me several months ago because she was coming out here for a conference – a writer's AGM. She's a writer. We emailed back and forth and then she came. We were in close connection. She invited me back to come and do an event there. She was able to find me through the Storytellers of Canada website.

Internet research, a Web profile on a networking website, an email address, and a few email exchanges helped to establish ties between these former strangers – ties that were later reinforced through in-person meetings and project collaboration.

For some cultural workers, email has been absorbed as a standard part of their audience outreach and marketing strategy. Email allows cultural workers to sustain their creative practice more effectively and economically by expanding their market reach into distant markets. An established professional painter in Surrey (3 July 2007) relies on email to connect with both art dealers and fans:

> I have dealers who are far away, but just ship to them and deal with them via email or phone calls. Emails seem to be an important part of the connection. I've been recently getting fan emails from obscure places in the US and Australia. I got one last night. This woman was very thankful for my painting, which I thought was nice. I think without email or the internet, one could be very isolated.

Email correspondence helps suburban cultural workers to combat a sense of isolation by connecting socially with friends, family, fans, colleagues, dealers, or clients.

Younger generations of cultural workers are particularly reliant on email and other social networking sites (e.g., MySpace and Facebook) to connect with collaborators, to learn about gigs, events, and performances, and to find outlets for their work. The Underground Garage Sale, a youth arts-and-crafts event in Mississauga was launched through email-driven networking. A visual artist in Mississauga (25 October 2006) who helps to organize security for the youth-driven event outlines how her daughter initiated the project:

> My daughter said to me: "Mum, the CNE is too big. I want to do something for the young artists in Mississauga because they would love to make small crafts and things and they don't get any opportunity to sell their work. I want to do an underground garage sale." She rented this church and she sold the tables for $20 to all those arts people. She sent out the emails and youth have a very strong networking capability. All the tables were sold. The first day we had 150 people who came for that. Second year, we had over 500 people. This year, *Mississauga News* sponsored it and Art Gallery of Mississauga sponsored it, and this thing is just out of proportion. It's an amazing event. Twenty dollars for a table, free entertainment, all these bands come and play for free and it's totally youth-oriented, a youth-organized event. We just look after it. We make sure that security is provided and children are safe. But the rest, the kids do it.

Email is the communication and networking medium through which younger creatives are comfortable and adept at initiating and planning events.

Another larger and more formal event on Mississauga's cultural calendar also relies heavily on email. The Muslim Fest is a two-day Islamic arts and entertainment festival that attracts more than 10,000 people and has over 250 volunteers. In an interview, the director (30 March 2007) emphasizes his organization's dependence on email to recruit participants, volunteers, and attendees: "Whether it's email or Facebook, we rely heavily on that. When we did our comedy show, we did not print a single sign. I mean the only thing we did was to rely on our partnerships with our other organizations." Partnerships with local mosques provides this organization with weekly advertising opportunities at Friday prayers and through email listservs, in return for a promotional table at the annual event. Providing resources in-kind is a common support strategy among non-profit arts organizations, a strategy that can be facilitated through email communication.

In a further example of efforts to build bridges with the local community in suburbia, the owner of a private gallery in Mississauga collaborated with the directors of the Blackwood Gallery on the University of Toronto Mississauga campus and the Art Gallery of Mississauga to initiate the Mississauga Art Hop. Email was essential in the planning of this five-hour event in which galleries are connected to each other by free bus rides, as this gallery owner (5 October 2007) documents:

> She [the director of the Art Gallery of Mississauga] didn't know me and I didn't know her, but I engaged her in conversation by email first and then she came to visit, and we agreed that we would do a gallery hop. Two or three years ago, we did the first gallery hop in which we had five galleries; four of them were public and we were the only private art gallery that participated. The following year, we had seven galleries, two other private galleries … this year, we're doing the gallery hop again and this year we've gotten Brampton involved.

Email correspondence has the potential to gather momentum and reach a wider audience in ways that written communication sent by "snail mail" does not. For the curator of A Month in Photography festival in Mississauga (11 November 2007) most of her "correspondence with artists happens over the internet. I post out a call to artists and that's done online … people take that information and then post it somewhere else. People pick up that information and then feed it along." With the "forward" function in email accounts, electronic messages and information posts can quickly reach a spatially dispersed and diverse audience.

All of these examples demonstrate that computer-facilitated communication through email is a common means by which suburban arts administrators and cultural workers mobilize participation in small- and large-scale cultural events and increase that participation over several event cycles. Email is affordable and accessible and permits its users to reach a wide audience quickly. In the above examples of events in Mississauga, email communication allowed organizers to raise the profile of their event, to stimulate local interest, and to attract more participants with each passing year.

A key mechanism through which email communication can be accelerated and more widely distributed to internet users is through the use of membership listservs. A listserv is an electronic mailing list that automatically sends email messages to individuals who are members of the list. Through listserv participation, suburban cultural workers in the research sample, many of whom are members of a range of different arts organizations (varying in their local, regional, national, and international reach), receive electronic newsletters and bulletins. E-newsletters and e-bulletins are important means of informal interaction and information sharing within cultural communities. They keep their members updated regularly about events and activities. They also help to extend the physicality of informal cultural spaces by increasing computer-facilitated social networking opportunities in the non-physical world of the internet. Through participation in listserv distribution lists, suburban cultural workers maintain broad, multi-stranded, and often supportive relationships across potentially substantial physical distances.

Investing in Computer-Facilitated Communication Infrastructure

Computer-facilitated information sharing is central to the production of culture in suburbia. The transfer of information and knowledge generated by diverse modes of learning and accessed through different social networks is made possible by high-speed broadband internet infrastructure. For the majority of suburban cultural workers in this research sample, individual access to "internet infrastructure reduces the feeling of psychological disparity and alienation" (Park, 2004, 284). The internet and the computer hardware and software that cultural workers use build new cultural and spatial configurations that connect people and places.

However, the inner and outer suburbs in this study are positioned differently in terms of the municipal investment in communication infrastructure and the recognition given to its importance to the cultural sector. Municipalities should be key to crafting policies in response to technological change (Hackler, 2006). Yet for the suburban municipalities studied in this book, information such as the location and capacity of local communication infrastructure is neither readily available in the public domain nor documented in cultural planning reports. On municipal websites such information is rarely addressed as a critical component of local infrastructure. Online, it is difficult to find any documentation about targeted technology infrastructure strategies, investment (by the public or the private sector) in wireless or fibre-optic networks, or statistics about the number of high-speed lines, free-standing wireless telecommunication towers, or satellite facilities. The lack of public information can hinder a municipality in competition between cities to attract new business investment. Hackler (2006) advocates that municipalities should assess and inventory their technology infrastructure, because it has become so central to economic growth. In Australia, forward-thinking policy makers have identified the need to bridge arts and culture policy with media and communications policy in recognition of "the growing importance of access to high-speed broadband as an enabler of participation in the cultural sphere" (Flew, 2012, 240). In the Canadian context, the importance of "fast, universal, and equitable digital access" seems to register only minimally in municipal cultural policy documents (Apperley et al., 2011, 1).

The City of Mississauga's official website announces that free wireless access zones are available at all public libraries, community centres, and arenas. The city also offers an online geomatics mapping service that can be used to map local capital projects, infrastructure, and landmarks. Sample capital projects listed on the website include road resurfacing, sidewalk and bicycle lane construction, intersection turn lane additions, and noise wall construction, but there is no mention of capital investments in communications infrastructure. In addition, the majority of landmark symbols relate to recreation, transportation, community, and cultural infrastructure. For example, there are map symbols available for bocce ball courts, boat launches, seniors centres, transit terminals, and school board offices, but there are no symbols for any aspect of broad-band network infrastructure.

The City of Surrey launched an online mapping service called "COSMOS" in 2005. It allows residents and businesses to access geospatial information provided by the city (www.surrey.ca/city-services/654.aspx). Among the layers of data available to map and view is information on properties (e.g., monuments, lot dimensions, legal plan boundaries, rights of way, and road allowances); community services (e.g., school catchments, garbage collection, fire hall zones); land use, parks and recreation, and transportation (e.g., transit and truck routes, roads, and greenways); and utilities (e.g., hydro, methane and gas lines, drainage, sanitation, and water). Despite the detail provided on municipal engineering projects, no information is publicly available on cultural or digital infrastructure. A decade ago, City of Surrey councillors did discuss the need to develop a telecommunications master plan. However, no such plan is yet listed on the city's website under its plans and strategies.[9]

The community infrastructure web page of the City of North Vancouver's official website reads: "Basic public services such as clean drinking water and an efficient road network are vital to creating a sustainable community," while "health and safety, accessibility and housing are also important aspects of community infrastructure and key components of the City's community planning work" (www.cnv.org). No mention is made of communication infrastructure. One of the few direct references to communications technologies on the website is the 2009 announcement that free Wi-Fi hotspots would be offered during business hours in the civic plaza near the main public library and the city hall. The City and District of North Vancouver are more progressive than Mississauga and Surrey in recognizing the importance of computer-facilitated communication technologies to cultural production.

A significant goal of North Vancouver's cultural planning initiative is to improve communication and information resources for the arts community on the North Shore. To this end, in the lead-up to the 2010 Winter Olympics, a marketing team was established with representation from community agencies to secure grant funding to hire a marketing consultant. The marketing plan rebranded an arts council magazine, *Arts Alive*, and its website in an effort "to let the people across the great divide know that we're [culturally] alive and healthy and vibrant" (arts administrator, North Vancouver, 29 May 2007). In her estimation the plan has worked: "The buzz, the noise, starts to reverberate and we've piqued interest in the North Shore. Five years ago nobody would have even noticed what was going on over here. They wouldn't even bother to look. But that has changed dramatically."

Building on the success of the *Arts Alive North Shore* promotion, the Arts Office, the bi-municipal agency of the City and District of North Vancouver that supports and promotes arts and culture, launched a website in the spring of 2010 with a new online cultural events calendar called *on!* (www.artsoffice.ca). The new website incorporates social networking capabilities for both creators and consumers of culture. The re-rebranding initiative was intended to create an easily recognizable name and website where artists, arts organizations, audiences, tourists, and municipal staff can obtain news, information, and resources. The online calendar allows cultural workers and arts organizations to upload event information for free as well as to showcase individual projects and artists. In the words of agency administrators, the website is envisaged as "an interactive meeting place that will launch users on a journey of discovery to find out what's happening, who's performing, how to get tickets, apply for funding, explore public art and more" (www.allianceforarts .com/blog/news-north-shore-launches-arts-office). It is perhaps too soon to gauge its success as a virtual social space for cultural workers. What is more important, however, is that politicians, cultural planners, and arts administrators in North Vancouver have taken steps to realize the potential of the internet as a networking and marketing tool to support suburban cultural production.

In many municipalities, urban and suburban alike, governments, businesses, and post-secondary educational institutions are frequently recognized as requiring high-capacity telecommunications fibre-optic cables, wire-lines and wireless infrastructure, but that same requirement is rarely officially acknowledged as central to the operation and stability of the cultural sector. This chapter emphasizes the importance of computer-facilitated communications technology infrastructure to the maintenance and vibrancy of the cultural economy in suburban locales. Apperley et al. (2011) reiterate that participation in the digital cultural economy demands consideration of the technical infrastructure (connection quality, download speeds); medium (wireless, copper, coaxial cable, optical fibre); connection (modem, antenna, gateway); service (DSL, Wi-Fi, satellite, and FTTP); and the social conditions of their use. It is important that municipalities assess the stock, use of, and possibilities for computer-facilitated communication infrastructure as a key component of cultural planning and mapping exercises in the twenty-first century.

Conclusion: Keeping Up with New Cultural Production Practices

The dramatic growth of suburban and post-suburban living since 1970 has required, and continues to require, the labour of cultural workers as the "new pioneers, new neighbours, new activists in the local and not-so-local networks of formal and informal associations" (Clapson, 2003, 146). Cultural workers are active and important agents in the suburbanization process because they initiate associative and imaginative impulses in suburban environments. Cultural production is both an individualistic and a collective activity, with a necessary associational dimension to it. Through participation in voluntary organizations and the initiation of new activities and events, cultural workers foster an organizational and creative ethos in suburbia. This suburban creativity is sustained across an expansive and fragmented geography by social networks formed through computer-facilitated communication. Through these networks, the size of suburbia is made manageable and malleable to local cultural workers. As forums of social interaction and mobilization, computer-facilitated communication networks support suburban cultural event planning, allowing "new forms of spacing" and "new kinds of social relation to exist" (Thrift 2006, 143). In the words of Arturo Escobar (2001, 167), "networks are the location of emergent local actors and the source of promising cultural practices and possibilities." Social networks contain the seeds of future creative opportunities.

Computer-facilitated communication increases the opportunities for encounter as well as for the free and quick sharing of ideas and skills among previously unassociated cultural workers. Although the social ties that cultural workers establish online through virtual communication may appear thinner and less durable than those established within more traditional communities, these ties may also be reinforced through occasional face-to-face contact. "Even virtual clusters of intangible information flows in cyberspace have a place-based component because they are directional in the electronic networks and represent intricate relations between information infrastructure and social groups in those places" (Park 2004, 283–4). Cyberspace commingles with physical space in that resources are used by suburban cultural workers in particular places (Crang, 2010). Thus suburban cultural practice is sustained by an oscillating movement between cyberculture on the internet and in other network-mediated spaces and cultural production and interventions in the physical spaces of the suburbs where cultural workers live and create.

To support future suburban cultural production, it is important that municipalities invest in computer-facilitated communication infrastructure and create virtual spaces where cultural workers can interact for mutual benefit. A consideration of the organizational capacity of suburban cultural workers and the quality of connections between them can help to make suburbs more emotionally and psychologically sustaining places for other residents (Landry, 2007). To this end, an assessment of the stock and use of computer-facilitated communication infrastructure should be a central component of suburban cultural planning and cultural mapping exercises.

Chapter Eight

Conclusion: Culture from the Outside In

What constitutes through division the "inner" and "outer" worlds of the subject is a border and boundary tenuously maintained for the purposes of social regulation and control. The boundary between inner and outer is confounded by those excremental passages in which the inner effectively becomes outer, and this excreting function becomes, as it were, the model by which other forms of identity formation are accomplished. In effect, this is the mode by which the Others become shit.

—(Butler, 1990, 134)

Unsurprisingly, suburbs are not fecal and neither is the cultural work that is produced there. Yet an (en)forced division between inner (urban) and outer (suburban) artistic and spatial worlds has, for much of the last half of the twentieth century, positioned suburban cultural producers and their products on the margins of cultural life, where they have been repeatedly denigrated within the art world, and in scholarly and popular imaginations. This book has dismantled these mischaracterizations of Canadian suburbia as acultural and uncreative. Through a preliminary exploration of the culturesheds of Toronto and Vancouver, this book has shown how detrimental it can be to the vitality and sustainability of a city-region's creative resources to perceive and invest in culture as a solely urban phenomenon. Because suburbs are not seen to conform physically or socially to the image of dense, clustered cultural districts is not to say that they are "'uncreative' zones – places of domestic consumption rather than sites of innovation, the arts and creativity" (Gibson & Brennan-Horley, 2006, 457). Like Gibson & Brennan-Horley (2006), I, too, have shown that to judge certain parts of a city-region as more creative than other parts is to disparage the political and cultural institutions of the latter and also the cultural workers who live and work there.

The story of Canadian suburbs written about here emphasizes that suburbs are not fixed categories of space. Suburbs are not static locations on the outer margins of cities, with measurable attributes that provide frozen settings for human activity (Pred, 1984). Instead, this book presents suburbs as places that are open and dynamic. Suburbs are described as places that are in a continuous state of change. They are "always under construction" (Massey, 2005, 9). They are "constantly in the process of 'becoming'" (Murdoch, 2006). This characterization of suburbs parallels ecological understandings of edge environments as transition zones characterized by turbulence, complexity, and a constant state of becoming (Doubleday, 1999).

The concept of becoming is part of a long-standing philosophical dualism with being. Where *being* can refer to "enduring essences and entities," *becoming* can refer to a "process unfolding over time" (Crang, 2005, 200) where something is made and remade. In describing an ontology of becoming, the anthropologist Alberto Jiménez (2003, 141) writes: "The world happens with us and, in choosing what set of practices we will enact and engage in, we are also choosing what world we want to live in. Through our engagements with and in the world, we become the spaces to which we have invested our practice." Jiménez reinforces how social and spatial becomings are intertwined for everyone, not just cultural workers. Although place is not a concept that is used by Deleuze and Guattari (1987), these authors do develop a philosophy of becoming focused on trajectories of how identities form and change. Processes of becoming are for Deleuze linked to flows of desire: raw experiences that pass through the body before meaning or cognition. Inspired by the language and the ideas found in the "intellectual toolbox" that is *A Thousand Plateaus* (1987), Kim Dovey (2010, 7) uses Deleuze and Guattari's philosophy of becoming to theorize "a conception of place as a territorialized assemblage, defined by connections rather than essences." Place, for Dovey (2010, 7), "is a dynamic ensemble of people and environment that is at once material and experiential, spatial, and social." His architecturally informed appreciation of place is certainly congruent with understandings of place forwarded in the scholarly geographical literature by Alan Pred (1984), Tim Cresswell (1996; 2004), and Doreen Massey (2005). In *For Space*, Massey (2005, 131) reiterates many of these ideas when she writes of "an understanding of place – as open, as woven together out of ongoing stories, as a moment within power-geometries, as a particular constellation within the wider topographies of space, and as in process, as unfinished business." The ontological insights of all these scholars on the nature of becoming and place provide a collaged conceptual base upon which to summarize the role of cultural workers in the current suburban place-making process.

The Culture of Place-Making in Suburbia

Professional cultural workers immerse themselves not only in their art but also in the places where they create and/or seek to reconstitute those places through their creative practice. The philosopher Edward Casey (1996, 18) argues that "to live is to live locally, and to know is first of all to know the place one is in." In this phenomenological tradition, humanistic geographers like Yi-Fu Tuan (1974) and Edward Relph (1976, 43) have also presented places as "profound centres of human existence" through which we perceive and experience the world around us. With the everyday routines and mundane practices of daily life comes an intense familiarity with, and emotional attachment to, the micro-geographies of the local. Using the metaphor of "place-ballet," David Seamon (1980) illustrates how places are performed daily through the movement and embodiment of people enacting their everyday lives. Places, then, should be understood as embodied relationships between people and the world around them; they are, Cresswell (2004, 37) reminds us, "constructed by people doing things and in this sense are never 'finished' but are constantly being performed" as events. An appreciation of places as "eventmental" (Casey, 2011), characterized by openness and change rather than boundedness and permanence, is apparent in the narratives suburban cultural workers tell about suburbia. Much like "places gather things, thoughts, and memories in particular configurations" (Escobar, 2001, 143), suburban cultural workers can be said to do the same through their creative labour. Cultural workers use the power of imagination to transform objects, ideas, recollections, and experiences into meaningful works of art that may acquire value and interpretive insight over time. Part introversion, part "extrospection," the creative process demands that cultural workers turn inwards for meaning and look outwards for inspiration; "from their perspective, new directions, modes of thinking and approaches to making are constantly being explored" (Egan & Macaulay, 2012). The diverse repertoire of skills employed in artistic and creative practice can also powerfully transform suburban topography physically and cognitively.

Although much of North American suburbia was originally conceived without any artistic motivation, this book has shown that creative intentions and interventions are overtly and covertly embedded in many contemporary suburbs. In Calgary, for example, the artistic interventions of the Arbour Lake Sghool explicitly use disorder and humour to transform a suburban cul-de-sac, yard, and home into counter-spaces of cultural production that resist social conventions and property norms. Through the gaze, the labour, and the cultural products of a new generation of suburban bohemians, the aesthetics of art are merged into the politics of suburban life. Such mergers offer important reminders of the need to appreciate use value over exchange value and to remember that the unstructured, in-between spaces of suburbia shelter many place complexities, intimacies, and possibilities.

Suburban cultural workers have been instrumental in creating sites and moments that challenge the dominant organization of suburbia through creative reinvention and adaptive reuse of space, but it is also important to acknowledge the conventionality of much individual locational decision making. This book has demonstrated that cultural workers are not so unusual in their suburban presence; they are often deeply embedded in the social and physical fabric of suburbia. In choosing to live and work in the suburbs, cultural workers highlight many of the same quality-of-life consumer preferences as the traditional Canadian middle-classes: homes that are affordable and spacious in family-oriented neighbourhoods that value closeness to nature, domestic convenience, freedom, and privacy. To those who own a car and can afford the price of a suburban home, such residential benefits allow the suburbs to be interpreted as creatively emancipatory spaces. But that emancipatory potential can be constrained by lack of formal, government-led investment in cultural infrastructure (particularly that which supports the process rather than the products of cultural labour) and by lack of public acknowledgment of the cultural work that occurs there. Paradoxically, such disinvestment, disregard, and invisibility can be both liberating and constraining. It offers opportunities (with limited financial and institutional support) to create with less competitive and critical scrutiny, yet few places where connections can be made with, and feedback obtained from, professional peers outside of the home location. Such infrastructural shortages have been the catalyst for the selective creative retrofitting of suburbia by cultural workers.

The organic emergence and flexibility of informal community cultural service hubs in Etobicoke, Mississauga, North Vancouver, and Surrey illustrate how cultural workers intervene in the suburbs to meet the need for grassroots workspaces for learning, ideas exchange, and collaboration. People with different disciplinary skills and backgrounds collaborate, learn from each other, and share knowledge and equipment.

This suburban model of cultural production consists of mutually supportive interactions and social and cultural relationships that are "intermittent, irregular, informal, and not based on contractual agreements," and allows creativity to happen in an "unplanned, haphazard, and flexible way" (Shorthose, 2004, 153). There is often an explicitly non-economic dimension to this cultural work. Motivations to produce it can be rooted in community development, educational, social justice, or art-for-art's-sake agendas that eschew market concerns in favour of social utility and this can make suburban cultural production difficult to measure in economic impact assessments (Shorthose, 2004). Despite the difficulty of quantifying the contributions that individual suburban cultural workers make to the cultural economies of city-regions, it is important to appreciate the range of often intangible positive externalities that creative labour contributes to the quality of life in the suburbs by "reinventing our communities for and with the citizens who live there" (Brault, 2010, 93).

While this book has described suburban cultural workers as quintessentially middle-class in their residential choices, it has also acknowledged that the relative conventionality of these choices does not preclude the creative place-making potential afforded by their presence. Suburban cultural workers' intense, lived proximity to the status quo is also coupled with an intellectual ability to simultaneously step outside of it to reimagine, to initiate critical debate about, and to make real changes to the suburban places and landscapes that they call home. Central to this transformative process are computer-facilitated communication networks that help to sustain a suburban creative tradition across an extensive, fractured, and potentially isolating geography. These complex and layered networks have become valuable sites of social interaction, mobilization, and local cultural-activity planning that find important material expression in the lived physical spaces of the suburbs.

In the production of culture in suburbia, cultural resources and distributed intelligence shift back and forth between cyberspace and suburban space. These shifting constellations of networked relationships help to animate suburban places. In artfully bridging virtual and material worlds, it is cultural workers as an occupational group, with their ground-up approach to innovation and sociability, who are well placed to be significant agents of change in suburbia. As creators, cultural workers "are adventurous, and take off without warning for territories not yet familiar to us, returning with words, images, movements, and sounds that fascinate, concern, question, disturb, reveal, fill us with wonder, or prepare us for changes in how we perceive things" as well as actual social and built fabric changes (Brault, 2010, 33–4). Thus cultural workers deserve to join the recognized list of influential professionals and civic leaders (politicians, policy makers, urban planners, developers, architects, financiers, and real estate agents) who play a documented role in transforming suburbs and in imagining suburban futures. Together with the many other suburban residents of all classes, ages, ethnicities, religions, and sexualities, cultural workers participate strongly in suburban social and physical transformation through an incremental process of change. In this way, suburbs continue to "evolve over time through gradual accretions," and the collective form comes to express "the imprint of a broad spectrum of interests" (Dunham-Jones & Williamson, 2009, 2).

Beyond the Flagship Blueprint: Practical Recommendations for Incremental Changes

The lesson to be learned from suburban cultural workers is the value of small-scale, incremental changes not only to suburban built form, but also to the adaptive reuses to which that built form can be put. Such synergistic adaptations are largely under-appreciated contributions to suburban place-making. They are valuable precisely because they augment citizen engagement in the public life of suburbs and they can foster support for the varied needs of an increasingly socially and culturally diverse resident base across an interdependent city-region.

The promise that suburban cultural workers hold as agents of change in suburbia needs to be underwritten by greater municipal support for cultural expression along the spectrum from amateur to professional, with particular concern for bridging generational, disciplinary, and cultural divides. It is the professional cultural worker who is most neglected in suburbia, with limited resources and minimal training made available. In planning for suburban infrastructure investment, greater emphasis needs to be placed on the production rather than the consumption side of the cultural sector equation. Particular attention should be directed to the creation of decentralized nodes of affordable cultural workspaces and to operational financing for arts and culture organizations. In place of further investment in cultural megaprojects and festivals (which are often financially inaccessible to local cultural workers), a greater impact could be made on the daily lives of suburban residents by focusing on the unspectacular. Gibson et al. (2012, 299) concur, emphasizing "that unheralded and prosaic sites of suburban creativity" such as "community halls, writers' centres, youth music studios, and art spaces … deserve better and more sustained financial support." In the suburban Canadian context, it is safety upgrades, retrofits, and maintenance needs of older buildings that require investment. Throughout suburbia many ageing, low-rise suburban buildings appear to lack the construction quality to merit full restoration yet have the potential to be retrofitted as informal cultural service hubs for local residents. Such cultural hubs need not be isolated and fragmented stand-alone projects. Instead, they could help to reduce spatial disparities by becoming vibrant nodes of community creativity for a spatially dispersed population in a polycentric metropolis.

Certainly, clustering is widely known to strengthen institutional ties and incubate interpersonal networks of trust and sociality. There is, however, no need to fixate on the creation of visible suburban cultural districts from scratch when in the suburbs "decentred and dispersed sites of creativity" are often "the norm rather than the exception" (Brennan-Horley, 2011, 11). Cultural work in the suburbs has a variegated geography, with practitioners connected to each other and to suburban and urban spaces in complex, but also ordinary and sometimes tenuous, ways (Brennan-Horley, 2011). Suburbs may lack the café and bar networking "buzz" of the central city, but there are many examples in this book where storefronts, strip malls, and schools have been retrofitted as community cultural service hubs. This revised geography of suburban cultural production needs to be recognized by urban cultural policy makers, who should nurture these modest and utilitarian "pre-existent 'germs'" of an existent local artistic scene (Mizzau & Montanari, 2008, 667). The sensible reminder to suburban municipal decision makers is to support and showcase local spaces and local talent, rather than to celebrate flagship venues and outside expertise.

That same advice can be applied productively to the hiring of suburban arts administrators. In order to strengthen the connections between suburban cultural institutions, organizations, and local arts audiences, it is important that arts managers live and work in the suburbs that they serve. In this way there is also greater potential to augment the mentorship of a more diverse array of local residents to better prepare them for future roles as cultural leaders in their respective suburban communities. Furthermore, if the creative talent of arts administrators and cultural workers at formative career stages is to be retained in the suburbs, cultural policy makers need to generate strategies to better connect individual creatives to each other and to city-regional, national, and global professional networks (Bennett, 2010).

Many suburban cultural activities often appear invisible to the uninformed. To all of the above recommendations, one might add increased media coverage and alternative outlets for communicating, advertising, and building public awareness of local cultural workers, cultural activities, and cultural spaces (Bennett, 2010). By augmenting internal and external awareness of suburban artistic and cultural scenes, ties of trust and collaboration are strengthened; in turn, these ties help to combat a sense of professional isolation and to anchor local scenes within wider cultural networks both inside and outside city-regions.

Suburbs can "play a major role in creative city economies" (Brennan-Horley, 2011, 1). To do so more effectively, however, municipal decision makers must acknowledge that significant numbers of cultural workers live and work in the suburbs, must recognize that suburban cultural work has a different geography from its urban counterpart, and must invest more substantially in suburban cultural infrastructure. If creative city policies continue to use an economic model that concentrates "resources among already culturally rich sections of the urban population, such policies run the risk of further accentuating spatial inequalities in the cultural realm" (Flew, 2011, 5).

It is time for Canadian cultural policy makers to engage in "whole-of-city thinking" (Luckman, 2009, 82). Attention needs to be directed to issues of cultural infrastructure equity between central cities and suburbs; in particular, to the provision of suburban cultural resources (e.g., money, services, and facilities) that would aid in the sustained "development of multi-skilled creative activities" (Shorthose & Strange, 2004, 56) that augment the "different aspirations for cultural expression, participation, and production" of amateurs, semi-professionals, and professionals (Evans & Foord, 2003, 179). Cultural policy makers and consultancy firms can no longer afford to generate policy from mechanistic economic-impact assessments or inner-city "one-size-fits-all" formulas (Collis et al., 2010) or city-centric "policy transfer" adaptations of successful ideas from elsewhere (O'Connor & Gu, 2010). "The paramount danger," Luckman et al. (2009, 72) assert, "is that an increasingly formulaic creative city agenda is imposed upon places in a damaging and/or unrealistic manner (Kong, 2000), running roughshod over local needs, aspirations, and already existing or vernacular creative expressions." Cultural policy, then, must be developed gradually and flexibly in response to the specific needs of different local constituencies and the material everyday realities of suburban cultural workers (Shorthose, 2004). Such an "'organic' approach to cultural policy" development is best supported by ethnographic qualitative research that can provide rich, detailed, and textured accounts of the complexities and situatedness of suburban cultural work. In so doing, it will give voice to the diverse and spatially embedded lived experiences of real people in different times and places (Shorthose, 2004, 157).

Canadian suburbs are at a critical juncture in policy- and place-making. In the current period of social and economic uncertainty, the Canadian suburban population continues to age and to be marked by tensions wrought by cultural differences and income disparities. New communications technologies have also contributed to significant changes in work, living, and consumption patterns. Suburbs are "the terrain in which to look toward the possibility of a new social contract" for sustainable, creative, and inclusive city-regions (Ingersol, 2006, 20). This book has revealed that suburban cultural workers can help to negotiate the terms and conditions of such a social contract by virtue of how effectively their socio-material practices uniquely texture, and work with and for, suburban places. The dynamic and mutually supportive relationships that suburban cultural workers continue to informally build between the arts and local communities need targeted municipal financial support and cultural policy recognition. Only then will the transformative vision of culture reconstituting the city-region from the outside in gain significant momentum.

The Arbour Lake Sghool

On the internet, the Calgary community of Arbour Lake is marketed to prospective residents with an emphasis on the fitness opportunities and active lifestyle that can be pursued in the foothills of the Rocky Mountains:

> Imagine living in an exclusive community that has everything you need within walking distance from your home. An area with a pristine lake brimming with rainbow trout, spectacular mountain views, neighbourhood parks and a regional bike and pathway system carved through rolling hills.
>
> The focal point of the community is a beautiful 10-acre lake set directly in the heart of the community; water so clean that you can see down 18 feet to the bottom. Enjoy swimming, boating, fishing, ice skating, BBQs, and much more. Your activities are only limited by your imagination – Arbour Lake offers a quality of life unmatched in Northwest Calgary (arbourlake. com).

Modelled 25 years ago by developer David Poppitt on upscale, lakefront communities in southern California, the first show homes in Arbour Lake were built in 1991, and the artificial lake was filled and opened for recreation in 1993. Although imagination is supposedly the only limitation to possibility, the lengthy list of community activities and facilities advertised on the website quite clearly does not extend beyond outdoor sporting pursuits to include arts and culture.

In Arbour Lake, as in many suburbs, the arts have not to been treated as a valued service by local decision makers. A Residence Association fee is levied annually from every property owner for daily lake operations and maintenance of the main entrances. Aside from a community hall, a YMCA, and a public library, investment in cultural infrastructure is limited. In the face of this situation, the Arbour Lake Sghool artist collective, having no local, purpose-built space to support cultural production and consumption, has reconfigured a detached single-family home into a studio, gallery, and an ever-changing work of art.

The Arbour Lake Sghool residence stands out rather than blends in. It was never tried on in a showroom. The house plan, along with finishings and upgrades, was not pre-selected from "a foreclosed organization of space" presented by a developer in artistic renderings and models (Kenyon, 2004, 35). The house is situated on a corner lot, set back from a street inhabited by predominantly white, middle-class families, where unfenced lawns and driveways make up the space between houses. The pastel blue home is unique on a street of beige, split-level, detached single-family houses. It was designed and built by the parents of two of the artists before the mortgage was turned over to their two sons. Its layout does not conform to the limited choices afforded in a developer's pattern book. Ownership has reconfigured the relationships of usage and meaning in this suburban space. As homeowners, the sons have relatively free rein within the city's by-law framework to modify and to use the home as they want. Three of the rooms have been rented to friends who studied fine art together at the University of Calgary, and together these five young men have formed a suburban artist collective. A broad network of friends (who themselves identify as athletes, farmers, musicians, comedians, and tradespeople, rather than artists), family members, neighbours, and strangers collaborate with the Arbour Lake Sghool to create ephemeral, site-specific, light-hearted performance and installation projects that rely on a sense of humour and play to "break the pattern of uniformity and shatter the tranquility of the cul-de-sac form" (www.thearbourlakesghool.com).

The cul-de-sac was a functional postwar suburban reinterpretation of the short dead-end street. It became popular in the postwar era for several practical reasons. First, the curvaceous street patterns inspired by the Garden Suburb urban planning movement left irregularly shaped parcels of land that had to be serviced by a street. Second, the curved street form was different from the pre-war standardized street grid and could be marketed by housing developers as symbolic of a new and better future (Martinson, 2000). The Arbour Lake Sghool, however, sees a different future for their cul-de-sac than that perhaps originally envisioned by developers, who prioritized cost-effectiveness and marketability over community building. As one artist collective member explains in an interview: "They watched us at the beginning of the year and thought 'great, here's another crazy project' and 'this is horrible' and 'our property values are going down' and then, almost uniformly, we've been supported, which has been a fantastic experiment and a great risk" (24 June 2008). This artist detects an underlying current of excitement in neighbourhood interactions and describes neighbours thanking them for changing the social dynamics and quality of life in the subdivision. In what follows, I briefly describe three of their more visible and controversial installations.

In 2004, the Arbour Lake Sghool undertook a performance project entitled *Trench Warfare*. Trenches were dug in the front lawn and for 24 hours the artists re-enacted First World War combat. Water balloons replaced guns and bombs, dry ice mimicked mustard gas, and hoses replaced machine gun turrets. This enactment of war was intended as a performative extension of the proliferation of themed suburban communities around Calgary's periphery that reduce history and culture into a marketable package.

In the spring of 2006, the Arbour Lake Sghool replaced their Kentucky Bluegrass suburban lawn with a crop of locally donated barley sewn and harvested by hand in an installation entitled *Harvest*. By growing a towering crop of barley, they created a landscape that contrasted sharply with a streetscape of garage-fronted homes, manicured lawns, and hard-surfaced asphalt streets and driveways. The city sent warnings about the height violation of their crop. The lawn they had grown "was deemed to be 'infested' with 'weeds' of an 'objectionable and unsightly nature forming a nuisance or fire hazard'" (Turner, 2008, F8). The artists responded with a crop harvesting and an impromptu block party where neighbours, friends, and artists learned how to scythe, flail, and winnow the barley. They appealed the fine at City Hall arguing that the barley crop was a xeriscape. The city ruled that growing barley is an agricultural pursuit that does not constitute a garden in the suburbs. Had the artists trimmed their lawn to the standard height and used chemicals to maintain it, likely the city would have had no quarrel with them. The installation became a piece of street theatre that inspired a replica installation in the basement of the Art Gallery of Calgary, titled *Grow Op*, and an instructional video, *How to Make Beer*, for a group exhibition at the Brampton Art Gallery in Ontario.

In 2005, the Arbour Lake Sghool constructed a two-and-a-half-storey cardboard mountainous volcano in the yard. Over the following month the cardboard tower gradually disintegrated and elicited several angry emails from neighbours to the parents of the current owners. One neighbour writes in a letter posted on the artist collective's website:

> Here are a few photos when the tower had some shape. It has been almost a
> month with the "structure." As I have said they are good neighbours gener-
> ally. They are not noise makers or wild party goers. We (not just us but sev-
> eral families) as neighbours find it difficult, as over the past year or so there
> have been many projects that would be perfectly acceptable on remote acre-
> age but in an area such as this, where everyone takes pride in their homes
> and gardens, makes us uncomfortable and quite frankly embarrassed as
> your house and yard stands out frequently as being remarkably different.
> Quite often when many guests, family, and visitors come by we are trying to
> explain the various belongings, artifacts, and events. What happens inside
> the house does not concern us and is none of our business, but when it
> comes outside it affects us and the look of the entire neighbourhood.
>
> I do not believe in just complaining but in having constructive sugges-
> tions. If your tenants/sons want to have outside events/projects/art work
> we sure would appreciate some prior consultation as to what is/will hap-
> pen, for how long and when we would see normalcy return to the outside of
> the house and yard areas. Should you want to build a high privacy fence and
> gate around the yard so they can have projects with less impact to the sur-
> rounding homes, I would gladly help build it. Your sons' consideration to
> their neighbours would be greatly appreciated. (www.thearbourlakesghool
> .com)

This letter highlights how some suburban residents cherish predictability, order, and privacy in their suburban dreamscape. The social art
projects of the Arbour Lake Sghool can be perceived as confrontational,
transgressing the realm of property and deliberately rupturing the relationship between expected and unexpected uses to which suburban
homes and yards can be put (Purves, 2005). As Amy Kenyon (2004, 55)
has argued in her book *Dreaming Suburbia*:

> The suburban dreamscape is one in which single-family houses are set back
> from the street, surrounded by well-kept lawns, but within sight of similar
> lawns and houses. The demand is for a series of detached spaces, which are
> nonetheless in contact with each other. Like-mindedness and community
> cooperation are the physical and emotional comforts of the subdivision.

The Arbour Lake Sghool interventions deliberately push the boundaries
of social acceptability in the visible exterior border spaces of the yard.

The artists refuse to reproduce unwritten "property rules that enjoin comportment, movement, and action and thus the 'self-restraint' associated with property" (Blomley, 2005, 283). They raise questions about the uses to which property can reasonably be put and the relationship between property, works of art, and the artistic subject (Lee, 2000). The works of art the Arbour Lake Sghool produce on the lawn are visual markers of difference and transgression in an interstitial space where the traditionally private realm spills over into the public realm (Blomley, 2005). At this point of detachment and connection to surrounding residential properties, some residents hold a sense of collective ownership over the visual dimensions of neighbouring yards and seek to regulate the perceived difference and disorder contained there.

Inside the home, the artists lay temporary claim to spaces and surfaces (e.g., light table, drawing table, kitchen counter) to realize different art projects. The living room has been converted into a gallery. The walls are painted white, track lighting installed, and a clean, empty space of possibility has been created. The carpets are a little less clean. Ghosts of stains remain, now transformed into works of art with colourful permanent marker outlines (*The Ancients*).

The windows of the home, traditionally a way to frame family life (Stamp, 2004), have also become temporary exhibition spaces. On the kitchen window, *Map Fragments* is an abstract black line drawing of suburban street forms. "The piece grew slowly in a viral, organic way; something like the real way in which suburbs expand across territory" (www.thearbourlakesghool.com). In other ground-floor windows the artists organize, arrange, and curate everyday found objects in unusual combinations, forming new associations and meanings from them. By "de-familiarizing the banal and creating a series of strange aestheticized artifacts" (www.thearbourlakesghool.com) the artists provoke questions about what suburban residents choose to display in their homes and why. They disrupt conventional consumer circuits of mass-produced decorative objects, challenging the way in which individual and family values are communicated through the ornamentation of the suburban home (Kwon, 1989).

In the fall of 2009, the Arbour Lake Sghool travelled to Toronto to intervene in the inner-suburban domestic landscape of Willowdale, North York, through an exhibition titled *The Leona Drive Project.* Organized by the Public Access Collective in collaboration with L.O.T.: Experiments in Urban Research Collective (www.l-o-t.ca), this exhibition ran over a 10-day period in October 2009. The academic curators, Janine Marchessault (York University) and Michael Prokopow (Ontario College of Art and

Design), obtained permission from the developer, Hyatt Homes, to temporarily use five vacant houses due to be demolished on Leona Drive as a gallery. Canadian artists who work in a variety of media (audio, architectural installation, landscape, projection, photography, sculpture, and performance) in conjunction with students from the Claude Watson Arts Program at Earl Haig Secondary School in North York created site-specific works of art exhibited in the interior and exterior spaces of these houses. The Arbour Lake Sghool constructed a miniature bungalow-style shack in the backyard of 17 Leona Drive, where they performed as squatters for the duration of the show. Their performative installation drew attention to different kinds of property ownership and occupancy, referencing some of the previous illegal-immigrant tenants who called these bungalows home. Peeling back the layers of social and architectural history on this residential site, other art projects referenced the sounds, colours, and textures of the lives of both imaginary and real former occupants. The goal of the exhibition was to "interrogate what has been lost from older identities and utopias, and investigate recent developments in suburbia where new patterns of community and conscience operate" (www.leonadrive.ca). Artistic interventions and exhibitions such as those by the Arbour Lake Sghool and the Public Access Collective explore the creative potentialities of the suburban domestic realm. Their short-term critical engagements offer valuable alternative imaginaries of suburbia.

Notes

1. Introduction

1 In a collaborative research project that examines the proportions and policy implications of Canadian suburbs, Gordon and Janzen (2011) have developed a series of 12 classification models to estimate the proportion of Canadians who live in suburban neighbourhoods in 33 census metropolitan areas and a sample of census agglomerations. Their models estimate that in 2006 suburban areas constituted approximately 80 per cent of the national metropolitan population and 66 per cent of the national population.

2 Australia has been described as a "commonwealth of suburban dwellers" and as "the world's 'first suburban nation'" (Stevenson, 1999, 213).

3 In the United States the centrifugal force of suburban demographic growth has been attributed to white, middle-class flight from central cities or economically under-performing regions (Kotkin, 2010b) and to immigrants settling in suburbs rather than in central cities (Tavernise & Gebeloff, 2010). Canada's suburban population increase, particularly in the three largest census metropolitan areas, is also partly associated with immigration.

4 In their *Atlas of Suburbanisms*, Markus Moos and Anna Kramer (2012) capture some of the richness and diversity of Canadian suburbs with dot density maps illustrating different variables (e.g., diversity, families, homes, travel and mobility, and work, education, and income) drawn from Statistics Canada census data for census metropolitan areas (http://env-blogs. uwaterloo.ca/atlas/). This atlas is a part of a Canadian SSHRC-funded Major Collaborative Research Initiative led by a colleague at the City Institute at York University, Roger Keil, entitled *Global Suburbanisms: Governance, Land, and Infrastructure in the 21st Century* (2010–17).

5   I am also inspired by Francesca Kay's (2009) description of sense of place in
her book *An Equal Stillness*, a story about the life of a fictional British woman
painter. Kay (2009, 112) writes: "Among … the poets and the artists, there
was often talk about the sense of place. About how vital such a sense was
to an artist, and the effect that place had on their work. But even after an
extravagance of conversation this sense stayed nebulous in concept, difficult
to define. Was it inborn, acquired, or wishful thinking only, a nostalgia for
the future? Did it exist in the artist's conscious mind or was it a projection
of the reader's or the viewer's? Without it was an artist lost for anchor?"

6   Norcliffe uses a case study of the comic book industry to illustrate how some
sectors of the cultural economy have witnessed "a substantial dispersion of
production to the suburbs, smaller towns, and even farther afield to the pro-
verbial 'electronic cottage'" (Norcliffe & Rendace, 2003, 243).

7   Journalist Marsha Lederman (2012, A4) questions whether the City of
Vancouver is "becoming a cultural backwater" because of a "cultural exodus." She documents how the Vancouver Playhouse Theatre Company,
along with a historic independent movie house, announced their closing,
and she interviews artists who discuss the challenges of maintaining liveli-
hoods with the high cost of living and lack of municipal and provincial support for arts and culture following the Olympics. The exodus of artists from
cities and their concomitant move to suburbs, smaller cities, and rural areas
can, however, be difficult to document statistically. Census data by occupation can be used to show employment declines in one place and increases in
other places, but it cannot provide the causal mechanisms for that change
or directly link the place of departure to the place of arrival.

8   The City of Port Moody in British Columbia, for example, strives to capitalize on the exodus of cultural workers from downtown Vancouver by branding itself the "City of the Arts."

9   While the film festival lifestyle is used to market the luxury condominiums
in the tower, $35 million of the $140 million in capital costs for the new
TIFF headquarters came from Premier Dalton McGuity's Liberal government. His government also agreed to classify TIFF as a "not-for-profit
arts training facility" so that the new facility could access $46 million in a
low-interest loan through the Ontario Strategic Infrastructure Financing
Authority, which was established four years ago as part of the ReNew
Ontario investment initiative (Adams, 2009, A5). The glitz, glamour, and
celebrity culture of the movie industry are the lure for international media
attention and the grease that lubricates the wheels of political and corporate financial support. Few other local arts training facilities have obtained
such ready access to cultural infrastructure money.

10 Toronto's cultural plan refers to Major Cultural Organizations in the city as "Majors" (e.g., the Art Gallery of Ontario, the Canadian Opera Company, the National Ballet of Canada, the National Ballet School, the Toronto Symphony Orchestra, the Gardiner Museum of Ceramic Art, Pride Toronto, and the Caribbean Cultural Committee/Caribana). Since the adoption of a cultural plan, these large institutional players in the cultural sector have received, on average, an increase in grants from the city of 8 per cent a year with the grant amount rising from approximately $3.3 million in 2003 to $4.4 million in 2007 (Culture Division, 2005 and 2008).

11 Frank Gehry's renovation of the Art Gallery of Ontario in Toronto came at a cost of $276 million and a loss of 29 jobs in 2009. The Ontario Public Service Employee Union (OPSEU) that represents AGO employees documents that since 2005 the number of full-time jobs has increased from 136 to 177 while the number of part-time jobs without benefits has increased from 159 to 266 (www.opseu.org/news/press2010/may-18-2010.htm).

12 The Cultural Capital of Canada program was initiated by the federal Department of Canadian Heritage in 2002. The purpose of the program is to support special activities that celebrate arts and culture in Canadian municipalities and "build a cultural legacy for the community by integrating arts and culture into overall community planning" (www.pch.gc.ca). Each year, municipalities of different sizes across the country apply for one of five awards and compete to receive designation as a Cultural Capital.

13 The nine arts occupational categories in the Canadian census under the "F0" category of "professional occupations in art and culture" include: producers, directors and choreographers; authors and writers; conductors, composers, and arrangers; actors and comedians; other performers; artisans and craftspersons; musicians and singers; visual artists; and dancers.

14 Hill and Capriotti (2009) note in their second report, *A Statistical Profile of Artists in Canada: Based on the 2006 Census*, that Statistics Canada data are an imperfect source for information about artists. They highlight a number of occupational categories that are not adequately addressed in Statistics Canada occupational classifications – filmmakers and media artists, for example, as well as the breadth of the writing profession.

15 When the cultural sector is considered more broadly to include heritage occupations (e.g., librarians, curators, and archivists) and cultural occupations (e.g., graphic designers, translators, and architects) with the nine arts occupations, the number of workers in Canada increases to 609,000 (3.3 per cent of the labour force) (Hill & Capriotti, 2009).

16 The names of potential participants were collected through online directories for municipalities, arts organizations, and galleries, through articles in

local newspapers and arts event flyers posted on community notice boards, and most significantly through word-of-mouth snowball sampling. Emails and phone calls were used to contact informants and to schedule interviews. The semi-structured interviews lasted anywhere between one and three hours. At the choice of participants, the interviews were held variously in homes, workplaces, and publicly accessible social spaces like coffee shops and restaurants. The interviews were recorded, fully transcribed, and coded using the qualitative data analysis software *NVivo*.

2. The Geography of Somewhere

1   The urban landscape historian Dolores Hayden (2003) in her book *Building Suburbia* has decoded the American suburbs as a cultural landscape. She documents seven vernacular patterns of suburban construction from the beginnings of borderland construction in 1820 up to 2000: picturesque enclaves (1850), streetcar build-outs (1870), mail-order and self-built sub-urbs (1900), mass-produced "sitcom" suburbs (1940), edge nodes (1960), and rural fringe intensification (1980). Each pattern, she maintains, is characterized by particular development practices, building technologies, marketing strategies, and architectural preferences. Intertwined in these suburban patterns are different class configurations and practices of racial segregation; suburbia is a place scarred by a history of economic, racial, eth-nic, and gender exclusions and expectations.

2   The name Parkdale was first used in 1875 by the Toronto House Building Association to distinguish this 50-acre residential subdivision from the neighbouring Irish working-class suburb of Brockton (Whitzman, 2009). The name of this model suburb was intended to signify both natural beauty and remoteness from the city. In the 1870s, when Parkdale was referred to as a "Flowery Suburb" it was a "haven of homeownership" to "industrial walk-to-workers as much as middle-class commuters" and was described as "clean, safe, thrifty, somehow chaste in its values and aspirations" (Whitzman, 2009, 198 and 101). Initially located on the margins of the city, by the 1920s and 1930s Parkdale became part of the older central city, engulfed by urban development on the north and west. Whitzman (2009) asserts that as Toronto's municipal government increasingly relied upon the rationality and spatial discipline of the urban planning profession the diver-sity of land uses and people (particularly the increasing number of woman-led households and non-British immigrants) came to be seen as a sign of urban decline.

3   Feminist critics have shown that the single-family suburban home was a place of isolation and confinement for women, disconnected from its

physical and social context. As Ross King (1996, 39) outlines, the suburban home, street, and neighbourhood with their "quiet, discreet, respectful surveillance" functioned as a new panopticon. "Women become their own keepers and the new bourgeois suburbia produces and reproduces the ideology of self-surveillance" (King, 1996, 39).

4   Richard Harris (1996) documents the important role played by "owner building" unimpeded by land developers or municipal governments, in the rise of unplanned blue-collar industrial suburbs near factories and other employment sites in Toronto in the first quarter of the twentieth century.

5   For example, "pods" are an area of single-use zoning; "LULU" is a locally unwanted land use; "strip" is an arterial road of car-oriented ribbon development; and "TOAD" is an acronym for a temporary, obsolete, abandoned, or derelict location.

6   Garreau defines edge cities as perceptually unified places possessing more than five million square feet of leasable office space, more than 600,000 square feet of leasable retail space, and thus having more jobs than bedrooms. Such a definition seeks to be quantifiable and precise. Other critics of urban form seek to convey more generally the changing shape and character of cities. In his description of the "hundred-mile city," Deyan Sudjic (1992), for example, strives to convey the incoherence, disorientation, selfishness, and fear that governs this new kind of car-oriented, American city.

7   According to Lang and Lefurgy's calculations, as of the 2000 United States census, there were 54 boomburbs with only four over 300,000 in population (Mesa, Arizona; Santa Ana and Anaheim, California; and Arlington, Texas). They use the term "baby boomburb" to refer to suburban jurisdictions with a population base between 50,000 and 100,000.

8   The municipalities of Mississauga and Surrey in the decade between 1976 and 1986 experienced a population growth of over 50 per cent. With a population of 374,005 in 1986, 544,382 in 1996, and 668,549 in 2006, Mississauga's population grew by 46 per cent between 1986 and 1996 and by 23 per cent between 1996 and 2006. Surrey experienced a similar dramatic expansion of its population base, growing 68 per cent between 1986 and 1996 (from 181,447 to 304,447) and 30 per cent between 1996 and 2006 (from 304,477 to 394, 976) (Statistics Canada).

9   A dead mall has been described by blogger enthusiasts as a shopping mall with high vacancy rates, non-traditional store occupancy, low consumer traffic levels, or dated and/or deteriorating infrastructure (www.deadmalls .com). Depending on the extent of death, the entire mall or areas of it may be sealed off from public access or be in a state of demolition or, occasionally, redevelopment. According to Tom Daniels (1999, 185), America has over 35,000 shopping malls, 4,000 of which could be classified as "dead."

10 Take, for example, a *Globe and Mail* article in which the journalist writes "A
woman was sexually assaulted for nearly five hours in a terrifying attack that
took place in an Etobicoke parking lot early Saturday morning, police said"
(Friesen, 2009, A10). An editorial in the *Toronto Star* by a doctoral student
at York University seeks to re-represent the "in-between city" as a place of
potential rather than despair; a place where activists, concerned parents,
urban entrepreneurs, young leaders, independent media outlets, and
community organizations "work tirelessly on issues of transit justice, tenant rights, and food security" in the "hope that one day their lives will not
include the drama of police raids, struggling schools, low wages, and long
commutes" (Black, 2010).

11 Certainly, in the United States, high unemployment and business bankrupt-
cies have eroded the tax base of many municipalities, forcing them to sell
assets to reduce deficits and maintain service. A prime example of this in
2009 was the sale of the Pontiac Silverdome in the suburban municipality
of Pontiac, Michigan, 50 kilometres north of Detroit. Built in 1975 at a cost
of $55.7 million, the 88,000-seat stadium was abandoned in 2002 when the
National Football League's Detroit Lions moved to a new downtown Detroit
facility. With annual maintenance costs of $1.5 million, city administrators
sold the stadium on a 127-acre parcel of land to a Toronto-based property
developer for $583,000. This sale provoked one journalist to speculate that
"it could prove difficult to entice a professional sports startup or major
artists to an aging facility in the suburbs" (Ladurantaye, 2009). Suburban
redevelopment initiatives can be a policy and funding challenge for North
American municipalities.

12 Many of Toronto's earliest suburbs that at the outset of their development
"mix[ed] industry, commerce, and various sizes of houses in the fine-
grained pattern of the central city" (Whitzman, 2009, 74), like Yorkville,
the Junction, and Parkdale, have been transformed in the twentieth
century into arts districts of cultural consumption and production (Bain,
2003). Such culture-led neighbourhood transformations, over an extended
time horizon, suggest interesting possibilities for contemporary inner
suburbs.

13 The National Film Board of Canada supported the creation of a web docu-
mentary titled *The Thousandth Tower: Stories from Inside a Toronto Highrise*
based on the lived experiences of newcomers residing in North Etobicoke
(http://highrise.nfb.ca). The photographic portraits reveal vertical residential structures surrounded by concrete and asphalt that isolate its residents
in rental units. Without safe parks and playgrounds and without furnished
apartment lobbies and functional community rooms, there are few inviting,

sacred, recreational, and community spaces where neighbours can connect with one another and build supportive social networks.

14 The journalist Dave LeBlanc (2012) documents how in Rexdale industrial units have become Sikh temples and specialty stores, former fast-food chains have been transformed into independent Indian restaurants, a flea market gives new retailers valuable experience, a shopping mall that lost its mainstream stores subdivided retail units to make them financially accessible to independent merchants, and a multiplex theatre now screens only Bollywood films.

15 First elected in 1978, and now in her twelfth term in office at age 92, Mayor McCallion is Canada's longest-serving mayor.

16 The boundaries of Metro Vancouver are the same as those used by Statistics Canada to define the Vancouver census metropolitan area.

17 Metro Vancouver has a governing board of directors that represent each municipality as appointed by each municipal council. Planning for regional parks, water and sewage, regional hospital planning, and strategic land use and transportation all come under the mandate of Metro Vancouver (www .metrovancouver.org).

18 In the first phase of the redevelopment project, a new waterfront walkway has been constructed from the base of Lonsdale Avenue to a 200-metre pedestrian-only pier. The second phase of the project introduces 15 buildings to the Burrard Dry Dock Pier and St Roch Dock site, combining residential towers, commercial office buildings, and a conference and hotel facility with a museum, galleries, and parks.

19 Cloverdale, for example, was developed as a farming community and retains a strong rodeo and agricultural tradition that is expressed in land-use patterns, the naming of streets and subdivisions, the commercial built fabric, the events sponsored, and the local art produced there.

3. (Un)usual Suburbanites?

1 Following Zukin's lead, David Cole (1987) examines three New Jersey cities close to Manhattan that became popular studio locations for artists displaced from New York City. From a consideration of the relationship between artists, galleries, and neighbourhoods, he determines that it is artists' living and working space that has the most profound effect on urban change. Cole's (1987, 391) research demonstrates that artists are "romanticized because of their willingness to live in run-down areas with old factories and warehouses or to break racial and ethnic barriers, and politicized because they displace low-income groups and initiate gentrification that

benefits land speculators, developers, realtors, and ultimately the upper middle class."

2　Feminist scholars have critically examined the gendered nature of the discourse on suburbanization. They have demonstrated how traditional postwar North American suburbia was a "domestic prison" (Friedan, 2001) underpinned by women's unpaid "invisible work" (Dyck, 1989). They have revealed how these "bourgeois utopias" (Fishman, 1987) have become harder to find as women's participation in the paid workforce has increased as a necessary means of maintaining careers and middle-class lifestyles in the face of rising costs of living (Rose & Villeneuve, 1998; Kern, 2005). And they have shown how women's labour force participation has created some contradictions between the "assumptions underpinning suburban communities and the reality of female residents' lives" (Strong-Boag et al., 1999, 178) (e.g., the isolation of stay-at-home mothers, the temporal-spatial constraints of working mothers who must manage competing household and work demands, and the shortage of child-care opportunities (Pratt, 2003)). Feminist research on suburbia has sought to document the active roles assumed by women in their negotiations, constructions, and experiences of suburban landscapes (Strong-Boag et al., 1999).

3　The moments that culminated in the production of *Splitting* – the cuts, the transformations, and the visits – were documented on film and the images later reworked by Matta-Clark into photographic collages, a book, and a short film.

4　Twoaddthree is a family art collective made up of two parents and their three young children. It was established in 2008 to coincide with the Liverpool 08 European Capital of Culture. Twoaddthree has a mandate to explicitly challenge the capitalism of culture. With a budget of 10 per cent of the family's annual net income (from salaries, freelance work, tax credits, and child benefits), the Institute for the Art and Practice of Dissent at Home has organized and hosted free events, residencies, discussions, interventions, performances, and exhibitions out of their council flat on a housing estate in Everton, just north of Liverpool's city centre. Their website is a significant tool for archiving their ongoing activities and connecting with other artists.

5　"The project, variously categorized as an architectural undertaking, as 'Schwitters' most important collage project,' or as a performatively elaborated sculptural program, entailed the ceaseless manipulation of the artist's Hanover studio. Stretching vertically and horizontally to adjacent rooms, it eventually resulted in an all-encompassing environment. The internal space was transformed by the aggregation of found materials, objects, and sculptural forms affixed to the architectural structure" (Mansoor, 2002, 1).

6 A tension between use value and exchange value is at the core of suburban history; residents want to enjoy suburbia while protecting their private property investment, and developers, land speculators, politicians, and other boosters of growth seek to generate profit from it (Hayden, 2003).

4. Why (Not) Here?

1 Preliminary ideas for this chapter were first explored in a book chapter written for an edited collection that examines spaces of vernacular creativity (Bain, 2009).

2 For Ley, the "new" middle class is a class fraction born out of the youth culture movements of the 1960s and is distinct from the "old" middle class. Where the traditional bourgeoisie was formed through family connections, attendance at élite secondary schools and universities, and the experience of high culture, the new middle class uses their social and cultural capital and hard work to get the education, jobs, and salary that support a middle-class lifestyle.

3 This count permits duplication, allowing one informant to identify more than one subcategory within a category.

4 In North America, the idea that pastoral environments can foster intellectual creativity dates back to the founding of American "urban" universities in the seventeenth century, many of which were modelled on the British tradition of locating campuses at an accessible remove from the central city (O'Mara, 2006).

5 The rise of the multicentric city-region has helped to transform the way in which North Americans think about the relationship between city and suburb. Joseph Rodriguez (1999) explains that in the nineteenth century the American city and suburb were thought to play distinct but equally important roles within a region – they were complimentary rather than conflictual environments. "The city dominated the regional economy and was the fount of culture, education, business, and industry. The suburb was less populous and more economically homogeneous than the city, and contained much less traffic and noise. The suburb provided a calm domestic environment and a reprieve from the hectic pace of the city. From the city came new ideas, inventions, and consumer products" (Rodriguez, 1999, 7). Some of this romanticism is still apparent in explanations that cultural workers provide for living and working in the suburbs. Yet their presence there explicitly challenges the notion that culture originates in the city rather than the suburb.

6 A national survey conducted by the Art Gallery of York University for a study entitled *Waging Culture: A Report on the Socio-Economic Status of Canadian*

Visual Artists (Maranda, 2009) determined that there are between 22,500 and 27,800 visual artists in Canada. In 2007, a typical visual artist with an average of six years of post-secondary education earned $20,000 from all income sources and, after expenses, lost money. Artists spent an average of 26 hours per week on their studio practice, 14.5 hours on art-related employment, and 7.6 hours on non-art-related employment. In addition, they volunteered approximately three hours a week at art-related activities. This report highlights the "significant economic penalty" that artists pay "to pursue their practice, and are, through real monetary contributions, replacing potential income-earning employment with what amounts to free labour" (Maranda, 2009, i).

7   There are parallels here to the findings of a study by Gibson et al. (2010) that examines how creative workers in the small city of Darwin in the sparsely populated Northern Territory of Australia experience both remoteness and proximity. In describing the pleasures and benefits of isolation, Gibson et al. (2010, 31) discovered that urban and suburban creative workers identified the colonial/frontier context, with its proximity to nature, Asia, and Aboriginal country, as a source of inspiration, highlighted the benefits of working independently, and emphasized the laissez-faire artistic freedom that distance from "metropolitan trends, fashions, and compulsions" permits.

8   Within the Australian context, Gordon Waitt and Chris Gibson (2009) have documented how Wollongong, a regional city of 280,000 about 80 kilometres along the coast from Sydney has become attractive to cultural workers. According to Waitt and Gibson (2009, 1233), this dormitory suburb has experienced sustained in-migration of certain creative workers who seek a lifestyle change and affordable housing in close proximity to "a market of global proportions." Cultural producers who can work from home-based studios have been attracted to a string of northern beachside suburban "villages" where there are older heritage homes close to the ocean, but few formal cultural facilities. "Understood as relatively close to Sydney, small, quaint, aesthetic, alternative and comparatively inexpensive, the northern suburbs, fashioned as discrete villages, have become home to increasing numbers of cultural workers connected to Sydney's creative industries (and to export markets via Sydney)" (Waitt & Gibson, 2009, 1243). In Canada, the cultural market, audience, professional development, and infrastructure possibilities of large cities like Toronto and Vancouver are certainly noted by suburban cultural workers, but they take a back seat to more conventional middle-class lifestyle considerations such as affordable home ownership and recreational access to nature.

9   But a sense of isolation is not a new phenomenon. Take, for example,
the experience of the Hungarian modernist artist Andor Weininger, who
studied under Wassily Kandinsky and Paul Klee at the Bauhaus school in
Germany before emigrating to Toronto in 1951 and leaving seven years later
for New York. Despite his Bauhaus training and the friendships he established with future members of Toronto's legendary Painters 11, Weininger
was unable to make inroads into the city's burgeoning modern art scene.
In his book *A Bauhausler in Canada: Andor Weininger in the '50s*, Oliver Botar
(2009) considers why this artist left so few traces on the Canadian artistic
imagination. One of the suggestions by a journalist reviewing this book was
that "simple geography contributed to Mr. Weininger's sense of isolation
and frustration" (Le Blanc, 2010, G4). When Weininger first emigrated
to Toronto he lived for four years in a two-storey apartment overlooking
Lake Ontario in Mimico, Etobicoke. Le Blanc (2010, G4) proposes that the
"apartment building was relatively isolated from the centre of the city which
may have contributed to the abstract artist's inability to succeed in Toronto."
It is to the isolation of an inner suburban live/work studio location that Le
Blanc directly attributes this artist's lack of career success.

5. Suburban Cultural Infrastructure Planning Formulas

1   A report released by the Canadian Federation of Municipalities puts the
cultural infrastructure deficit at $40.2 billion (Mirza, 2007). This figure
for community, recreational, cultural, and social infrastructure is nested
within a larger municipal public infrastructure deficit that is estimated at
$123 billion.
2   Of all cultural services it is only libraries that are listed as part of city services
together with winter snow and ice control operations, building and construction, drainage and environment, fire, police, and emergency services,
engineering and customer services, land development process, online services, property taxes and utilities, transportation, and waste collection.
3   Surrey hosts two high-profile cultural events: the Surrey International
Writers Conference and the Surrey Children's Festival.
4   The suburban municipality of Markham on the north-eastern border of
Toronto has recently completed the data collection portion of this exercise and has created a publicly available online cultural directory of local
cultural activities, businesses, and organizations (http://www.ylm.ca/ylm/
ylm_home.aspx?f=markhamcultural).
5   Toronto continues to look with competitive respect to Vancouver and
Montreal for the revenue-generating strategies implemented to fund their

cultural assets. Both Vancouver and Montreal raise money for cultural tourism through a hotel occupancy levy, and both have a higher per capita investment in culture than does Toronto. In 2001, the per capita investment in culture in Canada's three largest cities was $12.93 in Toronto, $17.71 in Vancouver, and $26.62 in Montreal (Culture Division, 2003, 55). In 2003, the per capita investment increased slightly in Toronto to $13.00 and increased substantially in Vancouver to $19.00 and Montreal to $32.00 (Culture Division, 2005). By way of North American comparison of per capita cultural investment figures in 2003, Chicago's was $16.00, New York City's $54.00, and San Francisco's $80.00 (Culture Division, 2005).

6   In 2002, Toronto initiated a Cultural Build Grants program to help with immediate physical infrastructure repairs.

7   In 1989, the City of North Vancouver and the District Municipality together with the District of West Vancouver established the Arts and Culture Commission of North Vancouver (formerly the North Shore Arts Commission), a tri-municipal arm's-length agency that provided grants administration and cultural planning to all three municipalities on the North Shore. When West Vancouver pulled out of this arrangement in 1990, the Arts Commission remained intact as a bi-municipal office serving the community along with cultural staff employed by the District Municipality, Arts Coordinators and Cultural Planners with the North Vancouver Recreational Commission, and the North Vancouver Community Arts Council. The overlap in cultural service delivery between these different agencies created confusion among councillors and members of the public, and spawned in 2006 an 18-month independent review process.

8   For three years, the Cultural Olympiad arts festival has funded different performances and exhibitions in Metro Vancouver. The final eight-week instalment leading up to the 2010 Winter Olympics featured 22 high-profile commissions in a program of 190 works, many of which were concentrated in downtown Vancouver and in proximity to Olympic venues. The director of the Cultural Olympiad, Robert Kerr, believed that the Olympics increased the profile of local art: "I think it turns people on to the vitality of the community … and people go, 'Oh my God, I had no idea we had this strength and this diversity within our cultural community'" (cited in Lederman, 2010, 19 January, R2). While the Olympics may provide a platform and some trickle-down benefits for some local emerging artists and venues, the majority of the headline acts were well-established cultural workers of some international renown in well-recognized cultural venues. The cultural legacy of the Olympics in the wider region may not be that pronounced.

9 The model for this program is Artsopolis, an online initiative developed by
the Arts Council of Silicon Valley, California, that provides a central platform for cultural events and links to ticket purchases on host websites (www
.livesv.com).

10 This observation is nicely illustrated by Michael McKinnie's (2007) study
of the role of the theatre industry in Toronto's urban development. His
examination of a number of Toronto's theatrical enterprises revealed that
the main theatre sites in the city, along with other key performance and
entertainment venues, are located within relatively well-defined downtown boundaries. A four-kilometre-by-four-kilometre district that extends
from Queen's Quay in the south to Dupont Street in the north, and from
Bathurst Street in the west to the Don River in the east contains 45 of the
55 performance venues identified by the Toronto Theatre Alliance as
occupied by its members or suitable for member productions (McKinnie,
2007). Eighty-two per cent of Toronto's main theatre venues, then, are
located in a relatively small downtown district; several of the remaining venues are clustered on its edges. This concentration is perhaps unsurprising
given that the area is well-served by public transportation and has the requisite density of restaurants, bars, and cafés.

11 Using a case study of Williamsburg, Brooklyn, Curran (2010) persuasively
argues that while advocates of clusters value proximity for its role in creativity and innovation, they ignore the displacement (what Hutton [2009]
might term "dislocation effects") of diverse manufacturing and working-
class economic actors with significant skills and innovations to contribute
to cities. Curran (2010, 883) concludes that "recognition of the industrial
sectors that continue to serve urban neighborhoods can potentially lead
to a more diverse, stable, and equitable city, for these businesses serve and
employ a truly diverse population, are strongly committed to place and tend
to pay more than their counterparts in the service sector."

12 Gunnar Törnquist and Ake Andersson offer two of the earliest interpreta-
tions of creative milieux. For Törnquist (1984), creative milieux evolve
over a long period and have four key components: information (to be
exchanged and traded), knowledge (in the form of bodies of work and
databases), activity competencies, and creativity. These four features are
said to combine to create new products, ideas, and processes. The longevity,
information, and networking dimensions of creative milieux as identified
by Törnquist are also important considerations for Andersson (1985).
Creative milieux, as understood by Andersson, are compact places in cities
that are culturally diverse, rich in knowledge, possess repositories of original
knowledge and skills, and are well connected through telecommunications

infrastructure. Such milieux are predicated upon sound financial bases and support for experimentation and transgression.

13 The term "cultural industry," in the singular, was first developed by Max Horkheimer and Theodor Adorno in 1937 to describe the processes and products of mass culture.

14 In the 1990s, in Europe, there was growing recognition of the economic significance of "creative industries" – industries that originate with individual creativity, talent, and skill and that can generate wealth by employing ideas and intellectual work (Hesmondhalgh, 2007). In Britain, the main creative industries subsectors broadly include the cultural industries, along with publishing, software development, engineering, advertising, higher education, and the operation of building-based facilities such as libraries, archives, museums, heritage sites, galleries, theatres, and concert halls (Montgomery, 2008).

15 Barcelona's Forum Park is a large cultural district located ten kilometres from the city centre on the waterfront. Its architectural centrepiece is the Forum Building, which contains an auditorium and exhibition hall that was designed by Swiss architects Jaques Herzog and Pierre de Meuron to host the inaugural international Universal Forum of Cultures event in 2004. Amsterdam's New Metropolis is a technology museum, exhibition space, and event venue designed by the Italian architect Renzo Piano and located at the entrance to a tunnel leading north out of the city. Rome's Parco della Musica is a large public music complex with three concert halls and an outdoor theatre designed by Renzo Piano on a 1960 Olympic site on the north side of the city.

16 A multiplex is a building, often a residence or cinema, that contains multiple units or component parts. When multiplex cinemas were first introduced in the 1960s they were large cinema complexes with more than five theatres where movies could be shown simultaneously and with multiple show times on different screens. Four decades later, megaplexes may have 20 or more screens and seating capacity for over 5,000 people.

17 The Shadbolt Centre for the Arts, formerly the Burnaby Arts Centre, opened in Deer Lake Park, Burnaby, in 1995 and has a year-round schedule of arts classes, performances, festivals, and exhibitions (www.shadboltcentre .com). With a particular focus on the performing, visual, and literary arts, the facility consists of two theatres, six dance studios, music rehearsal rooms, and pottery studios.

18 Twenty-eight hours a week of shared studio space can be rented at the LAC at a rate of $320 per month.

19 For example, the Muslim Fest, a festival celebrating Muslim culture that was first held in August 2004, received a user-fee reduction. As a consequence,

festival organizers maintain that the event has helped to introduce the
LAC to the Muslim community in Mississauga. In the words of one festival
organizer: "You have this building called the Living Arts Centre and nobody
would ever think of going there ... but you start having Islamic school graduation parties happening there, regular Friday prayers happening there,
and Islamic conferences and it becomes more accessible" (arts administrator, Mississauga, 8 November 2007). The growing Muslim community
in Mississauga is frequently characterized as young, but as the number of
second- and third-generation Muslims increases, so does participation and
investment in Islamic cultural events and entertainment. For many new
immigrant Muslim parents who may be concerned with providing their
children with faith-based education focused on prayer and reading of the
Koran, questions of creativity and outlets for creative expression may be less
of a priority.

6. The Creative Art of Place Complexity in Suburbia

1   Kevin Lynch developed new techniques for analysing and representing
    the perceptual structure of cities. In a book that has become a classic,
    Image of the City, Lynch (1962) explores people's perceptions of places and
    their mental maps and way-finding techniques. He argues that "success-
    ful" urban landscapes possess the qualities of legibility and imageability.
    Legibility refers to the organization of basic elements of a city (e.g., paths,
    edges, nodes, districts, and landmarks) that, if developed in an interrelated
    way, can create a sense of a coherent whole and can contribute to a posi-
    tive image of a place. Imageability refers to the ability of objects in a city to
    evoke a strong emotive response from an observer. For Lynch, a city that is
    particularly legible and imageable possesses buildings and areas that are dis-
    tinctive but also interconnected.
2   Through the eighteenth and nineteenth centuries, the bourgeoisie in many
    Western cultures attempted to distance themselves from the working classes
    and their places of work through the location and construction of their
    homes. As warehouses, workshops, and factories replaced domestic manu-
    facturing, "the house became for the first time the centre of family life" and
    "the hub of middle-class social organization" (Whitehand & Carr, 2001, 7).
3   For example, Manitoba, Saskatchewan, and Quebec have helped school
    boards with declining enrolments to transform schools into "community
    hubs" by housing additional services for children and their families (e.g.,
    dental clinics, mental health services, and public libraries, computer labs,
    and swimming pools) (Mahoney, 2008, A7). In Toronto's creative-city plan-
    ning framework, the phrase "community cultural hubs" is used occasionally

to refer to strategies to promote inclusion and cultural engagement among the city's young people (e.g., City of Toronto, 2008b, 17).

4 The Cultural Services Department of the City of Toronto identifies 11 bricks-and-mortar "cultural centres" under its jurisdictional supervision, four of which are located in the inner suburb of Etobicoke: Assembly Hall, Etobicoke Civic Centre Art Gallery, Franklin Carmichael Art Centre, and Neilson Park Creative Centre. (Of the remaining seven, three are located in Scarborough, one in North York, and the remaining three in downtown Toronto.) In reality, these centres are not particularly central, nor do they serve a diverse population base.

5 Each year, for example, Jarnail Arts hosts an end-of-year performance at the Bell Performing Arts Centre, the largest theatre in Metro Vancouver. Attached to the Sullivan Heights Secondary School and built by the Surrey School District in 2002 at a cost of $12 million, this 1,100-seat venue hosts numerous community-led as well as commercially promoted events for the South Asian community in Surrey. There appears to be very little cultural cross-over between the Indo-Canadian and the Anglo-Canadian communities through cultural events hosted in Surrey, as the audiences remain quite distinct and often use the performance space quite differently. The manager for the Bell Performing Arts Centre explained that he oversees a skeletal Anglo-Canadian staff, who, like the volunteers, know very little about Indian or Pakistani culture, do not speak Punjabi, and have little awareness of Indo-Canadian community press coverage. In an interview, the manager (Surrey, 18 May 2007), provided an interpetation of the main cultural differences in facilities usage by the South Asian community. Indo-Canadian events, particularly the larger pop shows organized by promoters, have tickets sold through local video stores rather than the box office, rarely start at the scheduled time, run for many hours with people coming and going for food served in the lobby, have a male-dominated audience that often drinks in the parking lot before the show, have seating allocated on a first-come-first-served basis, and require heavy security to manage the crowd. These racialized observations of cultural facility usage emphasize an ethnocultural divide in Surrey that is reinforced through the management structure of facilities and has yet to be overcome in practice.

6 The arts, as Hudnut (2003, 195) asserts, "strengthen town centers; make living in a community attractive as a long-term investment for families, individuals, and businesses; build bridges between different ethnic groups; and add ambience and vitality to a city, whether they take the form of painting, music, sculpture, or even food. Artists lease low-rent space; convert empty buildings into useful ones; establish studios, galleries, and retail outlets; and generate business that expands the tax base."

7   Kolb (2008, 73) defines historical density as "a kind of diachronic thickness in a place ... where social norms involve reference to a history that has been sedimented as people lived there and responded to it over time." Time is the variable that allows layers of experience, use, and meaning to accrue.

7. Networked Geographies of Suburban Cultural Workers

1   The gathering of artists, writers, and critics in cafes and bars to discuss cultural issues and to exchange information and ideas has been a common occurrence (particularly for men) in bohemian districts of large European and American cities since the mid-nineteenth century (Zakon, 1978). The best-known example in twentieth-century North America is the Cedar Street Tavern on University Place just above Eighth Street in New York City, to which the suburban artist quoted above makes reference. For over a decade, the first generation of postwar abstract expressionists socialized with critics and curators in the Cedar Street Tavern. This romantic idea of a group of legendary painters meeting at a bar to discuss art and life has served as a model for generations of artists that followed.

2   Based on attributes related to qualities of stability and governance, Grabher (2006) has developed a fourfold typology of networks that are commonly studied within economic sociology and economic geography: informal networks, strategic networks, project networks, and regional networks. The first three have the greatest potential applicability to suburban cultural production. According to Grabher (2006), informal networks are based on shared experience, reciprocal relationships, and interpersonal ties. Strategic networks are more centrally and hierarchically organized and are formed out of inter-organizational alliances. Finally, project networks are temporally limited and combine interpersonal and inter-organizational alliances. Suburban cultural workers and suburban arts administrators participate in various capacities in different networks, some of which could be described as informal and interpersonal; others have a more strategic and project orientation and involve interaction with different arts organizations.

3   In an effort "to appreciate the multiplicity of network logics and to elucidate the role of thin and ephemeral network ties" within project networks, Gernot Grabher collaborated with Oliver Ibert (2006, 253) on an empirical analysis of organizational project ecologies in the German advertising and software business. This collaboration produced a typology of personal knowledge networks that uses the degree of social embeddedness to distinguish between connectivity, sociality, and communality networks. For Grabher and Ibert (2006, 258), personal networks emerge out of many project cycles and can be "an untidy relational space" of potential distractions,

tensions, conflicts, and exploitations. Communality networks are characterized by deep, strong, and long-lasting relationships built on mutual loyalty
and trust. Sociality networks are more ephemeral, flexible, and strategically
career-oriented in nature, and involve ongoing investments of time, money,
and attention to keep in touch through casual and private forms of face-
to-face interaction. Finally, connectivity networks are constructed from weak
social ties formed from extended webs of virtual connection and exchange
and are used to solve problems and overcome skill-set deficiencies. This
typology was created within the research context of advertising and software
development industries and thus has a specific business- and project-driven
focus and application that do not neatly overlay the networking experiences
of suburban cultural workers. Nevertheless, the language used in the classification system is useful for its ability to convey the importance of a quest
for a sense of community, sociability, and connectivity in the social networks
that suburban cultural workers construct and participate in to help sustain
their creative practices.

4   Some cultural networks specific to particular immigrant groups are
    described by cultural workers (who are outsiders to those networks and
    ethnocultural groups) as exclusive, tightly knit, and self-sustaining.

5   Uniform resource locators are the global addresses of documents and
    resources available on the World Wide Web.

6   Bandwidth is a measurement in bits per second of the amount of data that
    can be transmitted over a network in a fixed time span.

7   It should be noted, however, that digital inequalities exist. Scholars have
    studied disparities in the distribution of information and communication
    technologies as well as the proficiencies to use these resources (Stern,
    2010). In a sociological study of young people's internet know-how,
    Hargittai (2010, 92) determined that "socioeconomic status is an important
    predictor of how people are incorporating the Web into their everyday lives
    with those from more privileged backgrounds using it in more informed
    ways for a larger number of activities."

8   In Mississauga, for example, the Beating Heart Festival's website is a
    MySpace page, which helps to reduce costs yet still allows organizers to post
    concert listings, publicize, and spread the word to other site users. Many
    local bands also make use of MySpace sites to quickly and freely advertise
    upcoming performances to fans.

9   There appears to be an ongoing lack of coordination among wireless com-
    munication service providers who negotiate separately with the city for
    licences, rights-of-way, and the option to erect free-standing wireless tele-
    communications towers.

References

Adams, J. (2009, 10 November). Ontario lightens TIFF Lightbox load. *Globe and Mail*, p. A15.

Amin, A., Massey, D., & Thrift, N. (2000). *Cities for the many not the few*. Bristol: Policy Press.

Andersson, A. (1985). Creativity and regional development. *Papers / Regional Science Association. Regional Science Association. Meeting*, *56*(1), 5–20. http://dx.doi.org/10.1007/BF01887900.

Apperley, T., Nansen, B., Arnold, M., & Wilken, R. (2011). Broadband in the burbs: NBN infrastructure, spectrum politics, and the digital home. *M/C 14*(4). http://journal.media-culture.org.au/index.php/mcjournal/article/view/400.

Artibise, A., & Meligrana, J. (2003). Tourism infrastructure of a postindustrial city: A case study of Vancouver, British Columbia. In D.R. Judd (Ed.), *The infrastructure of play: Building the tourist city* (pp. 271–295). London: M.E. Sharpe.

Artibise, A., & Meligrana, J. (2005). Regional governance and sustainability: The case of Vancouver. In Fritz Wagner, Timothy Joder, Anthony Mumphrey, Krishna Akundi, & Alan Artibise (Eds.), *Revitalizing the city: Strategies to contain sprawl and revive the core* (pp. 96–125). London and New York: M.E. Sharp.

Bader, I., & Scharenberg, A. (2010). The sound of Berlin: Subculture and the global music industry. *International Journal of Urban and Regional Research*, *34*(1), 76–91. http://dx.doi.org/10.1111/j.1468-2427.2009.00927.x.

Bain, A.L. (2003). Constructing contemporary artistic identities in Toronto neighbourhoods. *Canadian Geographer*, *47*(3), 303–317. http://dx.doi.org/10.1111/1541-0064.00024.

Bain, A.L. (2004a). In/visible geographies: Absence, emergence, presence, and the fine art of identity construction. *Tijdschrift voor Economische en Sociale Geografie, 95*(4), 419–426. http://dx.doi.org/10.1111/j.1467-9663 .2004.00318.x.

Bain, A.L. (2004b). Female artistic identity in place: The studio. *Social and Cultural Geography, 5*(2), 171–193. http://dx.doi.org/10.1080/ 14649360410001690204.

Bain, A. (2005). Constructing an artistic identity. *Work, Employment and Society, 19*(1), 25–46. http://dx.doi.org/10.1177/0950017005051280.

Bain, A.L. (2006). Resisting the creation of forgotten places: Artistic production in Toronto neighbourhoods. *Canadian Geographer, 50*(4), 417–431. http://dx.doi.org/10.1111/j.1541-0064.2006.00155.x.

Bain, A.L. (2007). Claiming space: Fatherhood and artistic practice. *Gender, Place and Culture, 14*(3), 249–265. http://dx.doi.org/10.1080/ 09663690701324870.

Bain, A.L. (2009). Creativ* suburbs: Cultural "popcorn" pioneering in Etobicoke. In T. Edensor, D. Leslie, S. Millington, & N. Rantisi (Eds.), *Spaces of vernacular creativity: Rethinking the cultural economy* (pp. 63–74). London, New York: Routledge.

Bain, A., & Marsh, J. (2012). Peterborough: A georegion in transition? In G. Nelson (Ed.), *Beyond the global city: Understanding and planning for the diversity of Ontario* (pp. 147–166). Montreal, Kingston: McGill-Queen's University Press.

Bain, A.L., & McLean, D. (2012). Eclectic creativity: Interdisciplinary creative alliances as informal cultural strategy. In B. van Heur & A. Lorentzen (Eds.), *Cultural political economy of small cities* (pp. 128–141). New York: Routledge.

Bain, A.L., & McLean, D. (2013). From post to poster to post-industrial: Cultural networks and eclectic creative practice. In M. Breitbart (Ed.), *Cultural economies in post-industrial cities: Creating a (different) scene* (pp. 97–121). Aldershot: Ashgate.

Bamber, B. (1997, April). From schoolhouse to studio. *Country Living, 136*: 88–95.

Barnes, T., & Hutton, T. (2009). Situating the new economy: Contingencies of regeneration and dislocation in Vancouver's inner city. *Urban Studies, 46*(5/6), 1247–1269. http://dx.doi.org/10.1177/0042098009103863.

Baumgartner, M.P. (1988). *The moral order of a suburb.* Oxford: Oxford University Press.

Beauregard, R. (2006). *When America became suburban.* Minneapolis: University of Minnesota Press.

Becker, C. (2009). *Thinking in place: Art, action and cultural production.* Boulder: Paradigm Publishers.

Becker, H. (1982). *Art worlds.* Berkeley: University of California Press.

Bell, D., & Jayne, M. (2010). The creative countryside: Policy and practice in the UK rural cultural economy. *Journal of Rural Studies, 26*(3), 209–218. http://dx.doi.org/10.1016/j.jrurstud.2010.01.001.

Bell, D., & Jayne, M. (Eds.). (2006). *Small Cities: Urban Experience beyond the Metropolis.* New York: Routledge.

Bennett, D. (2010). Creative migration: A Western Australian case study of creative artists. *Australian Geographer, 41*(1), 117–128. http://dx.doi.org/10.1080/00049180903535626.

Berelowitz, L. (2005). *Dream city: Vancouver and the global imagination.* Vancouver: Douglas and McIntyre.

Beuka, R. (2004). *SuburbiaNation: Reading suburban landscape in twentieth-century American fiction and film.* New York: Palgrave Macmillan.

Bianchini, F. (1993). *Urban cultural policy in Britain and Europe: Towards cultural planning.* Gold Coast: Institute for Cultural Policy Studies, Griffith University.

Bianchini, F., & Ghilardi, L. (2007). Thinking culturally about place. *Place Branding and Public Diplomacy, 3*(4), 280–286. http://dx.doi.org/10.1057/palgrave.pb.6000077.

Binnie, J. (1997). Coming out of geography: Towards a queer epistemology. *Environment and Planning D: Society and Space, 15*(2), 223–237. http://dx.doi.org/10.1068/d150223.

Bishop, P., & Williams, L. (2012). *The temporary city.* London, New York: Routledge.

Black, S. (2010, 14 May). A bad week in In-Between City. *Toronto Star.* http://www.thestar.com/opinion/editorialopinion/article/809027-a-bad-weck-in-in-between-city.

Blackwell, A. (2006). The gentrification of gentrification and other strategies of Toronto's creative class. *Fuse Magazine, 29*(1), 28–37.

Blomley, N. (2005). Flowers in the bathtub: Boundary crossings at the public-private divide. *Geoforum, 36*(3), 281–296. http://dx.doi.org/10.1016/j.geoforum.2004.08.005.

Boddy, T. (2007, 16 March). Exporting the "Vancouver style." *Globe and Mail,* p. S4.

Botar, O. (2009). *A Bauhausler in Canada: Andor Weininger in the '50s.* Oshawa: Gallery One One One.

Boudreau, J.-A., Keil, R., & Young, D. (2009). *Changing Toronto: Governing urban neoliberalism.* Toronto: University of Toronto Press.

Bourdieu, P. (1993). *The field of cultural production.* New York: Columbia University Press.

Bourne, L. (1996). Reinventing the suburbs: Old myths and new realities. *Progress in Planning, 46*(3), 163–184. http://dx.doi.org/10.1016/0305-9006(96)88868-4.

Bourne, L., Hutton, T., Shearmur, R., & Simmons, J. (Eds.). (2011). *Canadian urban regions: Trajectories of growth and change.* Oxford: Oxford University Press.

Bourriaud, Nicolas (2002). *Relational aesthetics.* Translated by Simon Pleasance and Fronza Woods. Paris: Les presses du réel.

Boyer, C. (1995). The great frame-up: Fantastic appearances in contemporary spatial politics. In H. Liggett & D. Perry (Eds.), *Spatial practices* (pp. 81–109). London: Sage.

Brault, S. (2010). *No culture, no future* (J. Kaplansky, Trans.). Toronto: Cormorant Books.

Brayshaw, C. (1998). *Edge city: New art from (& about) suburbia.* City of Surrey: Surrey Art Gallery.

Brennan-Horley, C. (2011). Reappraising the role of suburban workplaces in Darwin's creative economy. *M/C Journal 14*(4): 1–12. http://journal.media-culture.org.au/index.php/mcjournal/article/view/356.

Brennan-Horley, C., & Gibson, C. (2009). Where is creativity in the city? Integrating qualitative and GIS methods. *Environment and Planning A, 41*(11), 2595–2614. http://dx.doi.org/10.1068/a41406.

Bromberg, A. (2010). Creativity unbound: Cultivating the generative power of non-economic neighbourhood spaces. In T. Edensor, D. Leslie, S. Millington, & N.M. Rantisi (Eds.), *Spaces of vernacular creativity: Rethinking the cultural economy* (pp. 214–225). London, New York: Routledge.

Brooks, A., & Kushner, R. (2001). Cultural districts and urban development. *International Journal of Arts Management, 3*(2), 4–15.

Buck-Morss, S. (2000). *Dreamworld and catastrophe: The passing of mass utopia in east and west.* Cambridge, MA: MIT Press.

Bula, F. (2011). Putting a centre in the sprawl. *Globe and Mail,* 10 May. https://secure.globeadvisor.com/servlet/ArticleNews/story/gam/20110510/SRPROPERTY0105ATL.

Bunting, T., & Mitchell, C. (2001). Artists in rural locales: Market access, landscape appeal, and economic exigency. *Canadian Geographer, 45*(2), 268–284. http://dx.doi.org/10.1111/j.1541-0064.2001.tb01488.x.

Bunting, T., Walks, A., & Filion, P. (2004). The uneven geography of housing affordability stress in Canadian metropolitan areas. *Housing Studies, 19*(3), 361–393. http://dx.doi.org/10.1080/0267303042000204287.

Burnley, I., Murphy, P., & Jenner, A. (1997). Selecting suburbia: Residential relocation to outer Sydney. *Urban Studies, 34*(7), 1109–1127. http://dx.doi.org/10.1080/0042098975754.

Burns, B. (2011). Untold tales of the intra-suburban female. *M/C Journal, 14*(4), 1–7. http://journal.media-culture.org.au/index.php/mcjournal/article/view/398.

Buttimer, A. (Ed.) (1983). *Creativity in context.* Lund: The Royal University of Lund.

Casey, E. (1996). How to get from space to place in a fairly short stretch of time. In Steven Feld & Keith Baso (Eds.), *Sense of Place* (pp. 14–51). Sante Fe: School of American Research.

Casey, E. (2011). Do places have edges? A geophilosophical inquiry. In S. Daniels, D. DeLyser, J.N. Entrikin, & D. Richardson (Eds.), *Envisioning landscapes, making worlds: Geography and the humanities* (pp. 65–73). London, New York: Routledge.

Castells, M. (1996). *The information age: Economy, society, and culture. The rise of the network society.* Oxford: Blackwell.

Castells, M. (1997). *The power of identity.* Oxford: Blackwell.

Castells, M. (1998). *End of millenium.* Oxford: Blackwell.

Castells, M. (2000). Materials for an explanatory theory of the network society. *British Journal of Sociology, 51*(1), 5–24. http://dx.doi.org/10.1080/000713100358408.

Catungal, J.P., & Leslie, D. (2009a). Placing power in the creative city: Governmentalities and subjectivities in Liberty Village, Toronto. *Environment and Planning A. Society and Space, 41*(11), 2576–2594. http://dx.doi.org/10.1068/a41245.

Catungal, J.P., & Leslie, D. (2009b). Contesting the creative city: Race, nation, and multiculturalism. *Geoforum, 40*(5), 701–704. http://dx.doi.org/10.1016/j.geoforum.2009.05.005.

Chiras, D., & Wann, D. (2003). *Superbia! 31 ways to create sustainable neighbourhoods.* Gabriola Island, BC: New Society Publishers.

Chodikoff, I. (2008). Fringe benefits: Cosmopolitan dynamics of a multicultural city. In P. Poletto, P. Beesley, & C. Molnar (Eds.), *Ourtopias: Cities and the role of design* (pp. 41–52). Toronto: Riverside Architectural Press.

Christensen, J. (2008). *Big box reuse.* Cambridge, MA: MIT Press.

Chung, A. (2009, 7 October). Artists happily stuck in a rut: Ravaged roads deliver a hole of inspiration for Montrealers who turn the pits into poetry. *Toronto Star.*

City of Mississauga Culture Division. (2009). *Mississauga culture master plan.* Mississauga: Culture Division.

City of North Vancouver (2009a). *City of North Vancouver 2009 community profile release 1 – Data inventory*. City of North Vancouver. http://www.cnv.org// server.aspx?c=3&i=254.

City of North Vancouver (2009b). *City of North Vancouver 2009 community profile release 2 – Neighbourhoods*. City of North Vancouver. http://www.cnv.org// 'server.aspx?c=3&i=254.

City of Surrey (2009a). *City of Surrey labour force fact sheet*. Surrey: Planning and Development Department. http://www.surrey.ca/Doing+Business/ Population+and+Demographics/Demographic+Fact+Sheets.htm.

City of Surrey (2009b). *City of Surrey citizenship and immigration fact sheet*. Surrey: Planning and Development Department. http:// www.surrey.ca/Doing+Business/Population+and+Demographics/ Demographic+Fact+Sheets.htm.

City of Toronto. (2008a). *Agenda for prosperity*. Toronto: City of Toronto.

City of Toronto. (2008b). *Creative city planning framework: A supporting document to the agenda for prosperity: Prospectus for a great city*. Toronto: City of Toronto.

City of Toronto Culture Division. (2001). *The creative city: A workprint*. Toronto: Economic Development, Culture, and Tourism Department.

City of Toronto Culture Division. (2003). *Culture plan for the creative city*. Toronto: Economic Development, Culture, and Tourism Department.

City of Toronto Culture Division. (2005). *Culture plan progress report*. Toronto: Economic Development, Culture, and Tourism Department.

City of Toronto Culture Division. (2008). *Culture plan progress report II*. Toronto: Economic Development, Culture, and Tourism Department.

Clapson, M. (2003). *Suburban century: Social change and urban growth in England and the USA*. Oxford, New York: Berg.

Coe, N. (2000). The view from out west: Embeddedness, inter-personal relations, and the development of an indigenous film industry in Vancouver. *Geoforum, 31*(4), 391–407. http://dx.doi.org/10.1016/ S0016-7185(00)00005-1.

Cole, D. (1987). Artists and urban development. *Geographical Review, 77*(4), 391–407. http://dx.doi.org/10.2307/214280.

Collis, C., Felton, E., & Graham, P. (2010). Beyond the inner city: Real and imagined places in creative place policy and practice. *Information Society, 26*(2), 104–112. http://dx.doi.org/10.1080/01972240903562738.

Coupland, D. (2000). *City of glass: Douglas Coupland's Vancouver*. Vancouver: Douglas and McIntyre.

Crang, M. (2005). Time: Space. In P. Cloke & R. Johnston (Eds.), *Spaces of geographic thought* (pp. 199–220). London: Sage.

Crang, M. (2010). Cyberspace as the new public domain. In C. Kihato, M. Massoumi, B. Ruble, P. Supirós, & A. Garland (Eds.), *Urban diversity: Space, culture, and inclusive pluralism in cities worldwide* (pp. 99–122). Baltimore: Johns Hopkins University Press.

Cresswell, T. (1996). *In place/out of place: Geography, ideology, and transgression.* Minneapolis, MN: University of Minnesota Press.

Cresswell, T. (2004). *Place a short introduction.* Malden, MA: Blackwell.

Cronin, A., & Hetherington, K. (Eds.). (2008). *Consuming the entrepreneurial city: Image, memory, spectacle.* London, New York: Routledge.

Cultural Plan Steering Committee. (2002). *North Vancouver cultural plan.* City and District of North Vancouver: Cultural Plan Steering Committee.

Cumbers, A., Routledge, P., & Nativel, C. (2008). The entangled geographies of global justice networks. *Progress in Human Geography, 32*(2), 183–201. http://dx.doi.org/10.1177/0309132507084818.

Cunningham, G. (2000). The riddle of suburbia: Suburban fictions at the Victorian fin de siecle. In R. Webster (Ed.), *Expanding suburbia: Reviewing suburban narratives* (pp. 51–70). New York: Berghahn Books.

Curran, W. (2010) In defense of old industrial spaces: Manufacturing, creativity, and innovation in Williamsburg, Brooklyn. *International Journal of Urban and Regional Research, 34*(4), 871–885. http://dx.doi .org/10.1111/j.1468-2427.2010.00915.x.

Danaher, G. (2007). The region as performance space: A distinctive take on the creative industries. *Studies in Learning, Evaluation, Innovation and Development, 4*(1), 11–19.

Daniels, T. (1999). *When city and country collide: Managing growth in the metropolitan fringe.* Washington, DC: Island Press.

Dault, G.M. (2001). Sub-urbanity: The sound of no hand clapping. In *Suburbia: An exhibition examining the nature of suburbia and its effect on the contemporary psyche* (pp. 5–12). Peel: Art Gallery of Peel.

Davison, L. (Ed.). (2001). *Connections 4 artists' projects.* City of Surrey: Surrey Art Gallery.

Davison, L. (Ed.). (2002). *Journey.* City of Surrey: Surrey Art Gallery.

Davison, L. (Ed.). (2005). *Surrey seen: Surrey stories.* City of Surrey: Surrey Art Gallery.

Debord, G. (1994). *The society of the spectacle.* New York: Zone Books.

Deleuze, G., & Guattari, F. (1987). *A thousand plateaus: Capitalism and schizophrenia.* London: Athlone Press.

Denis-Jacob, J. (2012). Cultural industries in small-sized Canadian cities: Dream or reality? *Urban Studies, 49*(1), 97–114. http://dx.doi .org/10.1177/0042098011402235.

Domosh, M., & Seager, J. (2001). *Putting women in place: Feminist geographers make sense of the world.* New York, London: Guilford Press.

Doubleday, N. (1999). Arctic worlds and the geography of imagination. In A. Butimer & L. Wallin (Eds.), *Nature and identity in cross-cultural perspective* (pp. 189–198). London: Kluwer Academic Publishers.

Dovey, K. (2010). *Becoming places: Urbanism/architecture/identity/power.* London, New York: Routledge.

Dowling, R. (2000). Cultures of mothering and car use in suburban Sydney: A preliminary investigation. *Geoforum, 31*(3), 345–353. http://dx.doi.org/10.1016/S0016-7185(99)00048-2.

Drake, G. (2003). "This place gives me space": Place and creativity in the creative industries. *Geoforum, 34*(4), 511–524. http://dx.doi.org/10.1016/S0016-7185(03)00029-0.

Driver, F. (2001). *Geography militant: Cultures of exploration and empire.* Oxford: Blackwell.

Dunham-Jones, E., & Williamson, J. (2009). *Retrofitting suburbia: Urban design solutions for redesigning suburbs.* Hoboken, NJ: John Wiley and Sons.

Dunphy, K. (2009). *Developing and revitalizing rural communities through arts and creativity: Australia.* Vancouver: Creative City Network of Canada.

Duxbury, N. (Ed.). (2008). *Under construction: The state of cultural infrastructure in Canada.* Vancouver: Centre of Expertise on Culture and Communities, Simon Fraser University.

Dyck, I. (1989). Integrating home and wage workplace – women's daily lives in a canadian suburb. *Canadian Geographer, 33*(4), 329–341. http://dx.doi.org/10.1111/j.1541-0064.1989.tb00919.x.

Egan, M., & Macaulay, P. (2012). *LOOK out. Visual arts winter exhibition.* Toronto: Harbourfront Centre.

Emerson, J., & Emerson, S. (2002). A quintessential village: The story of Streetsville. In F. Dieterman (Ed.), *Mississauga: The first 10,000 years* (pp. 159–176). Toronto: eastendbooks.

Emirbayer, M., & Goodwin, J. (1994). Network analysis, culture, and the problem of agency. *American Journal of Sociology, 99*(6), 1411–1451. http://dx.doi.org/10.1086/230450.

ERA Architects Inc., Urban Intelligence Inc., and Guests Systems Inc. (2003). *A Map of Toronto's Cultural Facilities: A Cultural Facilities Analysis.* Toronto: Division of Economic Development, Culture, and Tourism.

Escobar, A. (2001). Culture sits in places: Reflections on globalism and subaltern strategies for localization. *Political Geography, 20*(2), 139–174. http://dx.doi.org/10.1016/S0962-6298(00)00064-0.

Evans, G. (2007). Branding the city of culture – The death of city planning? In J. Monclus & M. Guardia (Eds.), *Culture, urbanism, and planning* (pp. 197–213). Aldershot: Ashgate.

Evans, G. (2009). Creative cities, creative spaces, and urban policy. *Urban Studies, 46*(5/6), 1003–1040. http://dx.doi.org/10.1177/0042098009103853.

Evans, G., & Foord, J. (2003). Shaping the cultural landscape: Local regeneration effects. In M. Miles & T. Hall (Eds.), *Urban futures: Critical commentaries on shaping the city* (pp. 167–181). London: Routledge.

Evans, G., & Foord, J. (2008). Cultural mapping and sustainable communities: Planning for the arts revisited. *Cultural Trends, 17*(2), 65–96. http://dx.doi .org/10.1080/09548960802090634.

Farrell, M. (2001). *Collaborative circles: Friendship dynamics and creative work.* Chicago: University of Chicago Press.

Felton, E., Collis, C., & Graham, P. (2010). Making connections: Creative industries networks in outer-suburban locations. *Australian Geographer, 41*(1), 57–70. http://dx.doi.org/10.1080/00049180903535576.

Fiedler, Rob, and Addie, J.P. (2008). Canadian cities on the edge: Reassessing the Canadian suburb. Toronto. York University, CITY Institute. http://city .apps01.yorku.ca/wp-content/uploads/2011/05/file_Canadian_Cities_on_ the_Edge-1.pdf.

Fien, J. (2005). Beyond the city edge. In E. Charlesworth (Ed.), *Cityedge: Case studies in contemporary urbanism* (pp. 234–244). Amsterdam: Elsevier Architectural Press.

Fillion, K. (2008, 4 August). It's the truth of monogamy: We find meaning because there are limitations. *Maclean's, 121,* 30–31.

Filion, P., & Gad, G. (2006). Urban and suburban downtowns: Trajectories of growth and decline. In T. Bunting & P. Filion (Eds.), *Canadian cities in transition* (3rd ed., pp. 171–191). Toronto: Oxford University Press.

Filion, P., Olson, R., & Bunting, T. (2011). The transition from interventionism to neo-liberalism in the in-between city: The experience of the Toronto inner suburb. In D. Young, P. Wood, & R. Keil (Eds.), *In-Between Infrastructure: Urban Connectivity in an Age of Vulnerability* (pp. 181–196). Praxis (e)Press.

Fishman, R. (1987). *Bourgeois utopias: The rise and fall of suburbia.* New York: Basic Books.

Fleming, F., & Merullo, A. (Eds.). (2005). *The explorer's eye: First-hand accounts of adventure and exploration.* Weidenfeld and Nicolson.

Fleming, R.L. (2008). *The art of placemaking: Interpreting community through public art and urban design.* London, New York: Merrell.

Flew, T. (2010). Toward a cultural economic geography of creative industries and urban development. *Information Society, 26*(2), 85–91. http://dx.doi.org/10.1080/01972240903562704.

Flew, T. (2011). Right to the city, desire for the suburb? *M/C Journal, 14*(4), 1–7. http://journal.media-culture.org.au/index.php/mcjournal/article/view/398.

Flew, T. (2012). Creative suburbia: Rethinking urban cultural policy – the Australian case. *International Journal of Cultural Studies, 15*(3), 231–246. http://dx.doi.org/10.1177/1367877911433746.

Flew, T., Gibson, M., Collis, C., & Felton, E. (2012). Creative suburbia: Cultural research and suburban geographies. *International Journal of Cultural Studies, 15*(3), 199–203. http://dx.doi.org/10.1177/1367877911433755.

Florida, R. (2002). *The rise of the creative class: And how it is transforming work, leisure, community, and everyday life.* New York: Basic Books.

Florida, R. (2005). *The flight of the creative class.* New York: Harper Business.

Foord, J. (1999). Creative Hackney: Reflections on "hidden art." *Rising East: The Journal of East London Studies, 3,* 38–64.

Ford-Smith, H. (2001). Whose community? Whose art? The politics of reformulating community art. In *No frame around it: Process and outcome of the A Space community art biennale* (pp. 11–30). Toronto: A Space Gallery.

Fowle, K., & Larsen, L.B. (2005). Lunch hour: Art, community, administrative space, and unproductive activity. In T. Purves (Ed.), *What we want is free: Generosity and exchange in recent art* (pp. 17–26). New York: State University of New York Press.

Frey, W. (2004). The fading of city-suburb and metro-nonmetro distinctions in the United States. In A. Champion & G. Hugo (Eds.), *New forms of urbanization: Beyond the urban-rural dichotomy* (pp. 67–88). Aldershot: Ashgate.

Friedan, B. (2001). *The feminine mystique* (2nd ed.). New York: Norton.

Friedman, A. (2002). *Planning the new suburbia: Flexibility by design.* Vancouver: UBC Press.

Friesen, J. (2009, 14 December). Woman sexually assaulted for five hours. *Globe and Mail,* p. A10.

Friesen, J., Grant, K., & Paperny, A. (2010, 17 April). The alternative Toronto election. *Globe and Mail,* pp. M1, M5.

Frith, S. (1997). The suburban sensibility in British rock and pop. In R. Silverstone (Ed.), *Visions of suburbia* (pp. 269–279). London, New York: Routledge.

Furniss, E. (1999). *The burden of history: Colonialism and the frontier myth in a rural Canadian community.* Vancouver: University of British Columbia Press.

Garreau, J. (1991). *Edge city.* New York: Random House.

Gertler, M., Florida, R., & Vinodrai, T. (2002). *Competing on creativity: Placing Ontario's cities in a North American context.* Toronto: Ontario Ministry of Enterprise, Opportunity, and Innovation.

Gibson, C. (2002). Rural transformation and cultural industries: Popular music on the New South Wales Far North Coast. *Australian Geographical Studies, 40*(3), 337–356. http://dx.doi.org/10.1111/1467-8470.00184.

Gibson, C. (2010). Creative geographies: Tales from the margins. *Australian Geographer, 41*(1), 1–10. http://dx.doi.org/10.1080/00049180903535527.

Gibson, C. (Ed.). (2011). *Creativity in peripheral places: Redefining the creative industries.* London, New York: Routledge.

Gibson, C., & Brennan-Horley, C. (2006). Goodbye pram city: Beyond inner/outer zone binaries in creative city research. *Urban Policy and Research, 24*(4), 455–471. http://dx.doi.org/10.1080/08111140601035275.

Gibson, C., Brennan Horley, C., Laurenson, B., Riggs, N., Warren, A., Gallan, B., & Brown, H. (2012). Cool places, creative places? Community perceptions of cultural vitality in the suburbs. *International Journal of Cultural Studies, 15*(3), 287–302. http://dx.doi.org/10.1177/1367877911433750.

Gibson, C., & Connell, J. (2004). Cultural industry production in remote places: Indigenous popular music in Australia. In D. Power & A. Scott (Eds.), *The cultural industries and the production of culture* (pp. 243–358). London, New York: Routledge.

Gibson, C., Luckman, S., & Willoughby-Smith, J. (2010). Creativity without borders? Rethinking remoteness and proximity. *Australian Geographer, 41*(1), 25–38. http://dx.doi.org/10.1080/00049180903535543.

Gibson, M. (2012). *Bildung* in the 'burbs: Education for the suburban nation. *International Journal of Cultural Studies, 15*(3), 247–257. http://dx.doi.org/10.1177/1367877911433747.

Gibson, T. (2004). *Securing the spectacular city: The politics of revitalization and homelessness in downtown Seattle.* Lanham, MD: Lexington.

Gibson-Graham, J.K. (2006). *Postcapitalist politics.* Minneapolis: University of Minnesota Press.

Giles, J. (2004). *The parlour and the suburb: Domestic identities, class, femininity and modernity.* Oxford, New York: Berg.

Girling, C., & Helphand, K. (1994). *Yard-Street-Park: The design of suburban open space.* New York: John Wiley and Sons.

Goldbard, A. (2006). *New creative community: The art of cultural development.* Oakland: New Village Press.

Gordon, D., & Janzen, M. (2011). Suburban nation? Estimating the size of Canada's suburban population. Canadian Association of Geographers, annual meeting, Calgary, AB.

Gospodini, A. (2009). Post-industrial trajectories of Mediterranean European cities: The case of post-Olympic Athens. *Urban Studies, 46*(5–6), 1157–1186. http://dx.doi.org/10.1177/0042098009103859.

Grabher, G. (2006). Trading routes, bypasses, and risky intersections: Mapping the travels of "networks" between economic sociology and economic geography. *Progress in Human Geography, 30*(2), 163–189. http://dx.doi.org/10.1191/0309132506ph600oa.

Grabher, G., & Ibert, O. (2006). Bad company? The ambiguity of personal knowledge networks. *Journal of Economic Geography, 6*(3), 251–271. http://dx.doi.org/10.1093/jeg/lbi014.

Graham, S., & Marvin, S. (2001). *Splintering urbanism: Networked infrastructures, technological mobilities and the urban condition.* London, New York: Routledge. http://dx.doi.org/10.4324/9780203452202.

Greenstein, R., & Wiewel, W. (Eds.). (2000). *Urban-suburban interdependencies.* Cambridge, MA: Lincoln Institute of Land Policy.

Gruneau, R., & Whitson, D. (1993). *Hockey night in Canada: Sport, identities and cultural practices.* Toronto: Garamond.

Hackler, D. (2006). *Cities in the technology economy.* Armonk, NY: M.E. Sharpe.

Hall, M., & Lee, B. (2010). How diverse are US suburbs? *Urban Studies, 47*(1), 3–28. http://dx.doi.org/10.1177/0042098009346862

Hall, P. (1998). *Cities in civilization.* New York: Fromm International.

Halseth, G. (1996). Mapping residential redevelopment in a Canadian suburb. *Canadian Journal of Urban Research, 5,* 138–146.

Hargittai, E. (2010). Digital na(t)ives? Variation in Internet skills and uses among members of the "net generation." *Sociological Inquiry, 80*(1), 92–113. http://dx.doi.org/10.1111/j.1475-682X.2009.00317.x

Harris, R. (1996). *Unplanned suburbs: Toronto's American tragedy 1900 to 1950.* Baltimore: Johns Hopkins University Press.

Harris, R. (2004). *Creeping conformity: How Canada became suburban, 1900–1960.* Toronto: University of Toronto Press.

Harris, R. (2007, 25 April). The suburbs. *National Post,* p. A17.

Harris, R., & Forrester, D. (2003). The suburban origins of redlining: A Canadian case study, 1935–54. *Urban Studies, 40*(13), 2661–2686. http://dx.doi.org/10.1080/0042098032000146830.

Harris, R., & Lewis, R. (2001). The geography of North American cities and suburbs, 1990–1950: A new synthesis. *Journal of Urban History, 27*(3), 262–292. http://dx.doi.org/10.1177/009614420102700302.

Hayden, D. (1984). *Redesigning the American dream.* New York: Norton.

Hayden, D. (2003). *Building suburbia: Green fields and urban growth 1820–2000.* New York: Pantheon.

Hayden, D. (2004). *A field guide to sprawl.* New York: W.W. Norton.

Head, D. (2000). Poisoned minds: Suburbanites in post-war British fiction. In
    R. Webster (Ed.), *Expanding suburbia: Reviewing suburban narratives*
    (pp. 71–89). New York: Berghahn.
Heebels, B., & van Aalst, I. (2010). Creative clusters in Berlin: Entrepre-
    neurship and the quality of place in Prenzlauer Berg and Kreuzberg.
    Geografiska Annaler. Series B, Human Geography, 92(4), 347–363. http://
    dx.doi.org/10.1111/j.1468-0467.2010.00357.x
Hesmondhalgh, D. (2007). *The cultural industries* (2nd ed.). London: Sage.
Hill, K., & Capriotti, K. (2009). A statistical profile of artists in Canada: Based
    on the 2006 census. *Statistical Insights on the Arts, 7*(4), 1–44.
Hill Strategies Research Inc. (2006). Artists in large Canadian cities. *Statistical
    Insights on the Arts, 4*(4), 1–40.
Hubbard, P., Kitchen, R., Fuller, D., & Bartley, B. (Eds.). (2002). *Thinking
    geographically: Space, theory, and contemporary human geography*. London, New
    York: Continuum.
Hudnut, W.H. (2003). *Halfway to everywhere: A portrait of America's first-tier
    suburbs*. Washington, DC: Urban Land Institute.
Hughes, A. (2000). Retailers, knowledges, and changing commodity networks:
    The case of the cut-flower trade. *Geoforum, 31*(2), 175–190. http://dx.doi.org/
    10.1016/S0016-7185(99)00034-2
Hulchanski, D. (2007). The three cities within Toronto: Income polariza-
    tion among Toronto's neighbourhoods, 1970–2000. *Centre for Urban and
    Community Studies Research Bulletin, 41*(December), 1–12.
Hume, C. (1997). La vie bohème. *Toronto Star*, p. G3.
Hume, C. (2012, 19 January). Mississauga waking up to a new reality. *Toronto
    Star*, http://www.thestar.com/mobile/NEWS/article/1118316.
Huskins, B. (1999). "Tale of two cities": Boosterism and the imagination of
    community during the visit of Prince of Wales to Saint John and Halifax in
    1860. *Urban History Review 28*(1), 31–46.
Husler, K. (1997). Visual browsing: auto-flaneurs and roadside ads in the
    1950s. In P. Lang & T. Miller (Eds.), *Suburban discipline* (pp. 9–19). New
    York: Princeton Architectural Press.
Hutton, T. (2004a). The new economy of the inner city. *Cities (London,
    England), 21*(1), 89–108. http://dx.doi.org/10.1016/j.cities.2004.01.002.
Hutton, T. (2004b). Post-industrialism, post-modernism and the repro-
    duction of Vancouver's central area: Retheorising the 21st-century
    city. *Urban Studies, 41*(10), 1953–1982. http://dx.doi.org/10.1080/
    00420980420002056332.
Hutton, T. (2009). The inner city as site of cultural production *sui
    generis*: A review essay. *Geography Compass, 3*(2), 600–629. http://dx.doi
    .org/10.1111/j.1749-8198.2008.00201.x.

Hutton, T. (2012). Multilevel governance and urban development: A Vancouver case study. In M. Horak & R. Young (Eds.), *Sites of governance: Multilevel governance and policy making in Canada's big cities* (pp. 263–298). Montreal, Kingston: McGill-Queen's University Press.

Indergaard, M. (2009). What to make of New York's new economy: The politics of the creative field. *Urban Studies, 46*(5/6), 1063–1093. http://dx.doi.org/10.1177/0042098009103855.

Ingersol, R. (2006). *Sprawltown: Looking for the city on its edges.* New York: Princeton Architectural Press.

Jackson, K. (1985). *Crabgrass frontier: The suburbanization of the United States.* Oxford: Oxford University Press.

Jackson, S. (2011). *Social works: Performing art, supporting publics.* New York: Routledge.

Jiménez, A.C. (2003). On space as capacity. *The Journal of the Royal Anthropological Institute, 9*(1), 137–153. http://dx.doi.org/10.1111/1467-9655.t01-1-00008.

Jindrich, J. (2010). When the urban fringe is not suburban. *Geographical Review, 100*(1), 35–55. http://dx.doi.org/10.1111/j.1931-0846.2010.00005.x.

Johansson, M., & Kociatkiewicz, J. (2011). City festivals: Creativity and control in staged urban experiences. *European Urban and Regional Studies, 18*(4), 392–405. http://dx.doi.org/10.1177/0969776411407810.

Johnson, L. (1994). *Suburban dreaming: An interdisciplinary approach to Australian cities.* Geelong, Victoria: Deakin Universitiy Press.

Johnson, L. (2012). Creative suburbs? How women, design and technology renew Australian suburbs. *International Journal of Cultural Studies, 15*(3), 217–229. http://dx.doi.org/10.1177/1367877911433744.

Jonas, A., & Ward, K. (2007). Introduction to a debate on city-regions: New geographies of governance, democracy, and social reproduction. *International Journal of Urban and Regional Research, 31*(1), 169–178. http://dx.doi.org/10.1111/j.1468-2427.2007.00711.x.

Jurca, C. (2001). *White diaspora: The suburb and the twentieth-century American novel.* Princeton: Princeton University Press.

Juris, J. (2005). Social forums and their margins: Networking logics and the cultural politics of autonomous space. *Ephemera: Theory and Politics in Organization, 5*(2), 253–272.

Karttunen, S. (1998). How to identify artists? Defining the population for "status-of-the-artist" studies. *Poetics, 26*(1), 1–19. http://dx.doi.org/10.1016/S0304-422X(98)00007-2.

Kay, F. (2009). *An equal stillness.* London: Orion.

Keith, M. (2005). *After the cosmopolitan: Multicultural cities and the future of racism.* New York: Routledge.

Kendig, H. (1979). *New life for old suburbs: Post-war land use and housing in the Australian inner city.* Sydney: George Allen and Unwin.

Kennicott, P. (2009, 27 December). Best of the decade: Architecture. *Washington Post,* http://www.washingtonpost.com/wp-dyn/content/article/2009/12/24/AR2009122400116.html.

Kenyon, A. (2004). *Dreaming suburbia: Detroit and the production of postwar space and culture.* Detroit: Wayne State University Press.

Kern, L. (2005). In place and at home in the city: Connecting privilege, safety and belonging for women in Toronto. *Gender, Place and Culture, 12*(3), 357–377. http://dx.doi.org/10.1080/09663690500202590.

Kern, L. (2007). Reshaping the boundaries of public and private life: Gender, condominium development, and the neo-liberalization of urban living. *Urban Geography, 28*(7), 657–681. http://dx.doi.org/10.2747/0272-3638.28.7.657.

Kern, L. (2010). *Sex and the revitalized city.* Vancouver: UBC Press.

King, R. (1996). *Emancipating space: Geography, architecture, and urban design.* London, New York: Guilford.

Kleon, A. (2012). *Steal like an artist: 10 things nobody told you about being creative.* New York: Workman Publishing.

Kneight, K., & Schwarzman, M. (2005). *Beginner's guide to community-based arts.* Oakland: New Village Press.

Knox, P. (2005). Vulgaria: The re-enchantment of suburbia. *Opolis, 1*(2), 33–46.

Knox, P. (2008). *Metroburbia, USA.* New Brunswick, NJ: Rutgers University Press.

Kobayashi, E. (2007, 13 February). Toronto looks towards the 905 for culture: Brampton throws an artsy bash. *Toronto Star.*

Kolb, D. (2008). *Sprawling places.* Athens, London: University of Georgia Press.

Kolb, D. (2011). Many centers: Suburban habitus. *City, 15*(2), 155–166. http://dx.doi.org/10.1080/13604813.2011.568701.

Kong, L. (2000). Cultural policy in Singapore: Negotiating economic and socio-cultural agendas. *Geoforum, 31*(4), 409–424. http://dx.doi.org/10.1016/S0016-7185(00)00006-3.

Kotkin, J. (2000). *The new geography: How the digital revolution is reshaping the American landscape.* New York: Random House.

Kotkin, J. (2010a). Suburban nation, but urban policies. *Politico.* www.joelkotkin.com/content/00314-suburban-nation-urban-policies.

Kotkin, J. (2010b). *The next hundred million: America in 2050.* New York: Penguin.

Krätke, S. (2010). "Creative cities" and the rise of the dealer class: A critique of Richard Florida's approach to urban theory. *International Journal of Urban and Regional Research, 34*(4), 835–853. http://dx.doi .org/10.1111/j.1468-2427.2010.00939.x.

Kruse, K., & Sugrue, T. (2006). Introduction: The new suburban history. In K. Kruse & T. Sugrue (Eds.), *The new suburban history* (pp. 1–10). Chicago, London: University of Chicago Press.

Kunstler, J.H. (1993). *The geography of nowhere: The rise and decline of America's man-made landscape.* New York: Simon and Schuster.

Kwon, M. (1989). Suburban home life: Tracking the American dream. In *Suburban Home Life: Tracking the American Dream* (pp. 1–5). New York: Whitney Museum of American Art.

Ladurantaye, S. (2009, 24 November). 88,000 seats, one "surprising" winner. *Globe and Mail,* pp. B1, B9.

Landry, C. (2000). *The creative city.* London: Earthscan.

Landry, C. (2007). *The art of city making.* London: Earthscan.

Lang, R. (2003). *Edgeless cities: Exploring the elusive metropolis.* Washington, DC: Brookings Institution Press.

Lang, R., & Lefurgy, J. (2007). *Boomburbs: The rise of America's accidental cities.* Washington, DC: Brookings Institution Press.

Langford, B. (2000). Margins of the city: Towards a dialectic of suburban desire. In R. Phillips, D. Watt, & D. Shuttleton (Eds.), *De-centring sexualities: Politics and representation beyond the metropolis* (pp. 64–80). London: Routledge.

LeBlanc, D. (2010, 15 January). Toronto's one-and-only Bauhausler. *Globe and Mail,* p. G4.

LeBlanc, D. (2012, 4 May). A walk on the northwest fringe, with a side of curry. *Globe and Mail,* p. G4.

Lederman, M. (2010, 19 January). $20-million, 190 performances, 8 weeks. *Globe and Mail,* pp. R1, R2.

Lederman, M. (2012, 15 March). Curtain falling on Vancouver's cultural scene. *Globe and Mail,* p. A4.

Lee, P. (2000). *Object to be destroyed: The work of Gordon Matta-Clark.* Cambridge, MA: MIT Press.

Lefebvre, H. (1991). *The production of space.* London: Blackwell.

Leslie, D., & Reimer, S. (1999). Spatialising commodity chains. *Progress in Human Geography, 23*(3), 401–420.

Lewis, R. (2002). The industrial suburb is dead, long live the industrial slum: Suburbs and slums in Chicago and Montreal, 1850–1950.

Planning Perspectives, 17(2), 123–144. http://dx.doi.org/10.1080/02665430110111847.

Ley, D. (1988). Social upgrading in six Canadian inner cities. *Canadian Geographer, 32*(1), 31–45. http://dx.doi.org/10.1111/j.1541-0064.1988.tb00853.x.

Ley, D. (1993). Past elites and present gentry: Neighbourhoods of privilege in Canadian cities. In L. Bourne & D. Ley (Eds.), *The changing social geography of Canadian cities* (pp. 214–233). Montreal, Kingston: McGill-Queen's University Press.

Ley, D. (1996a). *The new middle class and the remaking of the central city.* Oxford: Oxford University Press.

Ley, D. (1996b). The new middle class in Canadian central cities. In J. Caulfield & L. Peake (Eds.), *City lives and city forms: Critical research and Canadian urbanism* (pp. 15–32). Toronto: University of Toronto Press.

Ley, D. (2003). Artists, aestheticisation and the field of gentrification. *Urban Studies, 40*(12), 2527–2544. http://dx.doi.org/10.1080/0042098032000136192.

Ley, D. (2010). *Millionaire migrants: Trans-Pacific life lines.* Malden, MA: Wiley-Blackwell.

Ley, D., & Dobson, C. (2008). Are there limits to gentrification? The contexts of impeded gentrification in Vancouver. *Urban Studies, 45*(12), 2471–2498. http://dx.doi.org/10.1177/0042098008097103.

Listokin, D., & Beaton, W.P. (1983). *Revitalizing the older suburb.* New Brunswick, NJ: Center for Urban Policy Research.

Lloyd, R. (2006). *Neo-Bohemia: Art and commerce in the postindustrial city.* London, New York: Routledge.

Lorentzen, A. (2009). Cities in the experience economy. *European Planning Studies, 17*(6), 829–845. http://dx.doi.org/10.1080/09654310902793986.

Lorinc, J. (2006). The city as cultureshed. In A. Wilcox, C. Palassio, & J. Dovercourt (Eds.), *The state of the arts: Living with culture in Toronto* (pp. 14–23). Toronto: Coach House Books.

Lorinc, J. (2008). The aging city: Can our urban spaces handle legions of retiring boomers? *The Walrus, 5*(1), 33–35.

Lowes, M. (2002). *Indy dreams and urban nightmares: Speed merchants, spectacle, and the struggle over public space in the world-class city.* Toronto: University of Toronto Press.

Luckman, S., Gibson, C., & Lea, T. (2009). Mosquitoes in the mix: How transferable is creative city thinking? *Singapore Journal of Tropical Geography, 30*(1), 70–85. http://dx.doi.org/10.1111/j.1467-9493.2008.00348.x.

Lucy, W., & Phillips, D. (2003). Suburbs: Patterns of growth and decline. In B. Katz & R. Lang (Eds.), *Redefining urban and suburban America: Evidence from Census 2000* (Vol. 1, pp. 117–136). Washington, D.C.: Brookings Institution Press.

Lucy, W.H., & Phillips, D.L. (2000). *Confronting suburban decline: Strategic planning for metropolitan renewal.* Washington, DC: Island Press.

Lynch, K. (1962). *Image of the city.* Cambridge, MA: MIT Press.

Machor, J.L. (1987). *Pastoral cities: Urban ideals and the symbolic landscape of America.* London: University of Wisconsin Press.

Mahoney, J. (2008, 23 May). Turning emptying schools into service hubs: Report. *Globe and Mail*, p. A7.

Mahtani, M. (2002). Tricking the border guards: Performing race. *Environment and Planning D: Society and Space, 20*(4), 425–440. http://dx.doi.org/10.1068/d261t.

Mansoor, J. (2002). Kurt Schwitters' *Merzbau*: The desiring house. *Invisible Culture: an Electronic Journal for Visual Culture, 4,* 1–14.

Maranda, M. (2009). *Waging culture: A report on the socio-economic status of Canadian visual artists.* Toronto: Art Gallery of York University.

Markusen, A. (2006). Urban development and the politics of a creative class: Evidence from a study of artists. *Environment and Planning A, 38*(10), 1921–1940. http://dx.doi.org/10.1068/a38179.

Markusen, A., & King, D. (2003). *The Artistic Dividend: The Hidden Contributions of the Arts to the Regional Economy.* Project on Regional and Industrial Economics, Humphrey Institute. Minneapolis: University of Minnesota.

Marsh, M. (1990). *Suburban lives.* New Brunswick, NJ: Rutgers University Press.

Marshall, A. (2000). *How cities work: Suburbs, sprawl and the roads not taken.* Austin: University of Texas Press.

Martí-Costa, M., & Pradel i Miquel, M. (2011). The knowledge city against urban creativity? Artists' workshops and urban regeneration in Barcelona. *European Urban and Regional Studies, 19*(1), 92–108. http://dx.doi.org/10.1177/0969776411422481.

Martinson, T. (2000). *American dreamscape: The pursuit of happiness in postwar suburbia.* New York: Carroll and Graf Publishers.

Massey, D. (2005). *For space.* London: Sage.

Mattingly, P.H. (1997). The suburban canon over time. In P. Lang & T. Miller (Eds.), *Suburban discipline* (pp. 39–51). New York: Princeton Architectural Press.

Mattingly, P.H. (2000). *Suburban landscapes: Culture and politics in a New York metropolitan community.* Baltimore, London: Johns Hopkins University Press.

Mayes, R. (2010). Postcards from somewhere: "Marginal" cultural production, creativity and community. *Australian Geographer, 41*(1), 11–23. http://dx.doi .org/10.1080/00049180903535535.

Mays, J.B. (2008, 12 September). In old Yorkville, a sameness that's beautiful. *Globe and Mail*, p. G2.

Mays, J.B. (2009, 18 September). Film fest's new home takes a bow. *Globe and Mail*, p. G2.

McAuliffe, C. (1994). Don't fence me in: Artists and suburbia in the 1960s. In S. Ferber, C. Healy, & C. McAuliffe (Eds.), *Beasts of suburbia: Reinterpreting cultures in Australian suburbs* (pp. 94–110). Melbourne: Melbourne University Press.

McGuinness, M. (2000). Geography matters? Whiteness and contemporary geography. *Area, 32*(2), 225–230. http://dx.doi.org/10.1111/j.1475-4762. 2000.tb00133.x.

McKinnie, M. (2007). *City stages: Theatre and urban space in a global city.* Toronto: University of Toronto Press.

McKnight Foundation (2002). *A new angle: Arts development in the suburbs.* Minneapolis, MN: McKnight Foundation. http://www.mcknight.org/ grant-programs/arts.

Menger, P.-M. (1999). Artistic labor markets and careers. *Annual Review of Sociology, 25*(1), 541–574. http://dx.doi.org/10.1146/annurev.soc.25.1.541.

Metro Vancouver. (2008a). *2006 Census Bulletin #10: Labour Force Activity.* http://www.metrovancouver.org/about/publications/Publications/Census_ Bulletin_10_LabourForce.pdf.

Metro Vancouver. (2008b). *2006 Census Bulletin #11: Jobs in Metro Vancouver.* http://www.metrovancouver.org/about/publications/Publications/2006_ Census_Bulletin_11_Jobs.pdf.

Miller, N., & Cavens, D. (2008). *City of North Vancouver 100 Year Sustainability Vision: GHG Measurement and Mapping, Technical Paper.* Vancouver: Design Centre for Sustainability.

Milroy, B.M. (2009). *Thinking planning and urbanism.* Vancouver: UBC Press.

Mirza, Saeed (2007). *Danger Ahead: The Coming Collapse of Canada's Municipal Infrastructure.* Ottawa: Federation of Canadian Municipalities.

Mitchell, C., Bunting, T., & Piccioni, M. (2004). Visual artists: Counter-urbanites in the Canadian countryside? *Canadian Geographer, 48*(2), 152–167. http://dx.doi.org/10.1111/j.0008-3658.2004.00053.x.

Mitchell, K. (1997). Different diasporas and the hype of hybridity. *Environment and Planning A, 15*(5), 533–553. http://dx.doi.org/10.1068/ d150533.

Mitchell, W. (2000). *E-topia*. Cambridge, MA: MIT Press.

Mizzau, L., & Montanari, F. (2008). Cultural districts and the challenge of authenticity: The case of Piedmont, Italy. *Journal of Economic Geography, 8*(5), 651–673. http://dx.doi.org/10.1093/jeg/lbn027.

Montgomery, J. (2008). *The newest wealth of cities: City dynamics and the fifth wave.* Aldershot: Ashgate.

Moore, M., & Prain, L. (2009). *Yarn bombing: The art of crochet and knit graffiti.* Vancouver: Arsenal Pulp Press.

Moos, M., & Kramer, A. (2012). *Atlas of suburbanisms*. Waterloo, ON: University of Waterloo. http://env-blogs.uwaterloo.ca/atlas/.

Moradi, G. (2007). Letter to the editor. *Fuse Magazine, 30*(2), 7.

Morris, D. (2012, 21 April). Keep your club. We'll be partying in the pews. *Globe and Mail*, p. M3.

Mumford, L. (1961). *The city in history: Its origins, its transformations, and its prospects.* New York: Harcourt Press.

Municipality of Metropolitan Toronto. (1994). *Metro culture plan: Redefining our cultural framework.* Toronto: Kramer Design Associates.

Murdoch, J. (2006). *Post-structuralist geography: A guide to relational space.* London: Sage.

Newman, H.K. (1999). *Southern hospitality: Tourism and the growth of Atlanta.* Tuscaloosa: University of Alabama Press.

Nicolaides, B. (2006). How hell moved from the city to the suburbs: Urban scholars and changing perceptions of authentic community. In K. Kruse & T. Sugrue (Eds.), *The new suburban history* (pp. 80–98). Chicago, London: University of Chicago Press.

Noble, M. (2009). *Lovely spaces in unknown places: Creative city building in Toronto's inner suburbs.* Research Paper 217. Toronto: Cities Centre, University of Toronto.

Norcliffe, G., & Rendace, O. (2003). New geographies of comic book production in North America: The new artisan, distancing, and the periodic social economy. *Economic Geography, 79*(3), 241–263. http://dx.doi.org/10.1111/j.1944-8287.2003.tb00211.x.

Oakley, K. (2009). Getting out of place: The mobile creative class takes on the local. In L. Kong & J. O'Connor (Eds.), *Creative economies, creative cities: Asian-European perspectives* (pp. 121–134). Berlin: Springer Media. http://dx.doi.org/10.1007/978-1-4020-9949-6_8.

Oakley, K. (2004). Not so cool Britannia: The role of the creative industries in economic development. *International Journal of Cultural Studies, 7*(1), 67–77. http://dx.doi.org/10.1177/1367877904040606.

O'Connor, J., & Gu, X. (2010). Developing a creative cluster in a post-industrial city: CIDS and Manchester. *Information Society, 26*(2), 124–136. http://dx.doi.org/10.1080/01972240903562787.

O'Mara, M. (2006). Uncovering the city in the suburb: Cold war politics, scientific elites, and high-tech spaces. In K. Kruse & T. Sugrue (Eds.), *The new suburban history* (pp. 57–79). Chicago, London: University of Chicago Press.

Oldenburg, R. (1991). *The great good place.* New York: Paragon House.

Oswell, D. (2002). Suburban tales: Television, masculinity and textual geographies. In D. Bell & A. Haddour (Eds.), *City visions* (pp. 73–90). Toronto: Pearson Education.

Park, S.O. (2004). Knowledge, networks and regional development in the periphery in the internet era. *Progress in Human Geography, 28*(3), 283–286. http://dx.doi.org/10.1191/0309132504ph491xx.

Parlette, V., & Cowen, D. (2011). Dead malls: Suburban activism, local spaces, global logistics. *International Journal of Urban and Regional Research, 35*(4), 794–811. http://dx.doi.org/10.1111/j.1468-2427.2010.00992.x.

Peck, J. (2011a). Neoliberal suburbanism: Frontier space. *Urban Geography, 32*(6), 884–919. http://dx.doi.org/10.2747/0272-3638.32.6.884.

Peck, J. (2011b). Creative moments. Working culture, through municipal socialism and neoliberal urbanism. In E. McCann & K. Ward (Eds.), *Mobile urbanism: Cities and policymaking in the global age* (pp. 41–66). Minneapolis: University of Minnesota Press.

Peck, J. (2005). Struggling with the creative class. *International Journal of Urban and Regional Research, 29*(4), 740–770. http://dx.doi.org/10.1111/j.1468-2427.2005.00620.x.

Penfold, S. (2004). Parking lots, drive-ins, and the critique of progress in Toronto's suburbs, 1965–1975. *Urban History Review. Revue d'Histoire Urbaine, 33*, 8–23.

Perl, Jed (2001, 22 January). The adolescent city. *The New Republic*, 23–30.

Peterson, I. (1999, 5 December). Some perched in ivory tower gain rosier view of suburbs. *New York Times*, pp. 1, 46.

Phelps, N. (2012). The sub-creative economy of the suburbs in question. *International Journal of Cultural Studies, 15*(3), 259–271. http://dx.doi.org/10.1177/1367877911433748.

Phelps, N., Parsons, N., Ballas, D., & Dowling, A. (2006). *Post-suburban Europe: Planning and politics at the margins of Europe's capital cities.* London: Palgrave Macmillan.

Phillips, R. (2004). Artful business: Using the arts for community economic development. *Community Development Journal, 39*(2), 112–122. http://dx.doi.org/10.1093/cdj/39.2.112.

Pickerill, J. (2003). *Cyberprotest: Environmental activism online.* Manchester, NY: Manchester University Press.

Podkalicka, A. (2011). To Brunswick and beyond: A geography of creative and social participation for marginalized youth. *M/C Journal, 14*(4), 1–9. http://journal.media-culture.org.au/index.phph/mcjournal/article/view/367.

Podmore, J. (1998). (Re)reading the "loft living" *habitus* in Montréal's inner city. *International Journal of Urban and Regional Research, 22*(2), 283–302. http://dx.doi.org/10.1111/1468-2427.00140.

Powell, D. (1993). *Out west: Perceptions of Sydney's western suburbs.* Melbourne: Allen and Unwin.

Power, D., & Scott, A. (Eds.). (2004). *Cultural industries and the production of culture.* London, New York: Routledge. http://dx.doi.org/10.4324/9780203392263_chapter_1.

Pratt, G. (2003). Valuing childcare: Troubles in suburbia. *Antipode, 35*(3), 581–602. http://dx.doi.org/10.1111/1467-8330.00340.

Pred, A. (1984). Place as historically contingent process: Structuration and the time-geography of becoming places. *Annals of the Association of American Geographers, 74*(2), 279–297.

Preston, V. (2002). A community enriched: Moving to Mississauga. In F. Dieterman (Ed.), *Mississauga: The first 10,000 years* (pp. 209–228). Toronto: eastendbooks.

Professional Environmental Recreation Consultants Ltd. (2008). *Parks, recreation, and culture ten-year strategic plan.* Surrey: Department of Parks, Recreation, and Culture.

Purves, T. (Ed.). (2005). *What we want is free: Generosity and exchange in recent art.* New York: State University of New York Press.

Quinn, B. (2005). Arts festivals and the city. *Urban Studies, 42*(5), 927–943. http://dx.doi.org/10.1080/00420980500107250.

Rahamimoff, A. (2005). Jerusalem: Lessons from a shared city. In E. Charlesworth (Ed.), *Cityedge: Case studies in contemporary urbanism* (pp. 68–83). Amsterdam: Elsevier Architectural Press.

Ray, B., Halseth, G., & Johnson, B. (1997). The changing face of the suburbs: Issues of ethnicity and residential change in suburban Vancouver. *International Journal of Urban and Regional Research, 21*(1), 75–99. http://dx.doi.org/10.1111/1468-2427.00059.

Regional Culture Committee (2009). Regular meeting agenda for October 14. Greater Vancouver Regional District. http://www.metrovancouver.org/

boards/Regional%20Cultural%20Committee/Regional_Culture_
October14_2009-Agenda.pdf.

Relph, E. (1976). *Place and placelessness*. London: Pion.

Relph, E. (1997). Sense of place. In Susan Hanson (Ed.), *10 Geographic Ideas That Changed The World* (pp. 17–39). New Brunswick, NJ: Rutgers University Press.

Resonance Creative Consulting Partners. (2008). *A City Creative to Its Bones: Envisioning Neighbourhoods Arts Hubs*. A discussion paper. Toronto: Toronto Arts Council Foundation.

Robinson, M. (2012, 6 July). Mississauga puts its arts scene on the map. *Globe and Mail*. http://license.icopyright.net/user/viewFreeUse.act?fuid= MTYONjQ5N%3D.

Rochon, Lisa (2007, 17 November). Could Mississauga ever, ever be sexy? *Globe and Mail*, p. R4.

Rodriguez, J.A. (1999). *City against suburb: The cultural wars in an American metropolis*. Westport, CT: Praeger.

Roodhouse, S. (2006). *Cultural qu4rters: Principles and practices*. Bristol: Intellect.

Roodhouse, S. (2009). Understanding cultural quarters in branded cities. In S.H. Donald, F. Kofman, & C. Kevin (Eds.), *Branding cities: Cosmopolitanism, parochialism, and social change* (pp. 75–88). London, New York: Routledge.

Rose, D., & Villeneuve, P. (1998). Engendering class in the metropolitan city: Occupational pairings and income disparities among two-earner couples. *Urban Geography, 19*(2), 123–159. http://dx.doi.org/10.2747/0272-3638 .19.2.123.

Russo, A., & Butler, D. (2006). *Cultural planning toolkit*. Vancouver: Creative City Network of Canada.

Seamon, D. (1980). Body-subject, time-space routines, and place ballets. In A. Buttimer & D. Seamon (Eds.), *The human experience of space and place* (pp. 148–165). London: Croom Helm.

Schmidt, B. (2005). Causing a scene. *Bookninja Magazine*. http://www .bookninja.com/?page_id=257.

Sewell, J. (2009). *The shape of the suburbs: Understanding Toronto's sprawl*. Toronto: University of Toronto Press.

Sharpe, W., & Wallock, L. (1997). Bold new city or built up 'burb? Redefining contemporary suburbia. In R. Mohl (Ed.), *The making of urban America* (2nd ed., pp. 309–331). Wilmington, DE: Scholarly Resources.

Short, J.R. (1999). Urban imagineers: Boosterism and the representation of cities. In A. Jonas & D. Wilson (Eds.), *The urban growth machine: Critical perspectives two decades later* (pp. 37–54). Albany, NY: State University of New York Press.

Shorthose, J. (2004). Accounting for independent creativity in the new cultural economy. *Media International Australia, 112,* 150–161.

Shorthose, J., & Strange, G. (2004). The new cultural economy, the artist and the social configuration of autonomy. *Capital and Class, 84,* 43–59. http://www.metaether.org/words/articles/articles/the%20new%20cultural%20economy,%20the%20artist%20and%20the%20social%20configuration%20of%20autonomy.pdf.

Sibley, D. (2001). The binary city. *Urban Studies, 38*(2), 239–250. http://dx.doi.org/10.1080/00420980125629.

Sies, M. (2001). North American suburbs, 1880–1950: Cultural and social reconsiderations. *Journal of Urban History, 27*(3), 313–346. http://dx.doi.org/10.1177/009614420102700304.

Silverstone, R. (Ed.). (1997). *Visions of suburbia.* London, New York: Routledge.

Simosen, K. (2008). Practice, narrative and the "multicultural city": A Copenhagen case. *European Urban and Regional Studies, 15*(2), 145–158. http://dx.doi.org/10.1177/0969776407087547.

Slater, T. (2004). Municipally managed gentrification in South Parkdale, Toronto. *Canadian Geographer, 48*(3), 303–325. http://dx.doi.org/10.1111/j.0008-3658.2004.00062.x.

Slater, T. (2005). Toronto's South Parkdale neighbourhood: A brief history of development, disinvestment, and gentrification. University of Toronto Centre for Urban and Community Studies. *Research Bulletin, 28,* 1–6.

Smith, A., & Strand, I.K. (2011). Oslo's new opera house: Cultural flagship, regeneration tool, or destination icon? *European Urban and Regional Studies, 18*(1), 93–110. http://dx.doi.org/10.1177/0969776410382595.

Smith, H. (2003). Planning, policy and polarization in Vancouver's Downtown Eastside. *Tijdschrift voor Economische en Sociale Geografie, 94*(4), 496–509. http://dx.doi.org/10.1111/1467-9663.00276.

Smith, N. (1996). *The new urban frontier: Gentrification and the revanchist city.* London, New York: Routledge.

Smith, R.G. (2003). World city actor-networks. *Progress in Human Geography, 27*(1), 25–44. http://dx.doi.org/10.1191/0309132503ph411oa.

Social Planning Council of Metropolitan Toronto. (1979). *Planning agenda for the eighties – part II: Metro's suburbs in transition.* Toronto: Social Planning Council of Metropolitan Toronto.

Soja, E. (1996). *Thirdspace: Journeys to Los Angeles and other real-and-imagined places.* Oxford: Blackwell.

Soja, E. (2000). *Postmetropolis: Critical studies of cities and regions.* Oxford: Blackwell.

Soja, E. (2008). Exopolis: The restructuring of urban form. In *Herbert Wright instant cities* (pp. 148–155). London: Black Dog.

Solnit, R., & Schwarzenberg, S. (2000). *Hollow city: The siege of San Francisco and the crisis of American urbanism.* London: Verso.

Sorensen, T. (2009). Creativity in rural development: An Australian response to Florida (or a view from the fringe). *International Journal of Foresight and Innovation Policy, 5*(1–3), 24–43. http://dx.doi.org/10.1504/IJFIP.2009.022097.

Sowden, T. (1994). Streets of discontent: Artists and suburbia in the 1950s. In S. Ferber, C. Healy, & C. McAuliffe (Eds.), *Beasts of suburbia: Reinterpreting cultures in Australian suburbs* (pp. 76–93). Melbourne: Melbourne University Press.

Stamp, R. (2004). *Suburban modern: Postwar dreams in Calgary.* Nanoose Bay, BC: Touchwood Editions.

Stewart, S. (2006). *Cultural mapping toolkit.* Vancouver: Creative City Network of Canada.

Stanilov, K., & Scheer, B.C. (Eds.). (2004). *Suburban form: An international perspective.* New York: Routledge.

Stanwick, S., & Flores, J. (2007). *Design city: Toronto.* Chichester: John Wiley & Sons.

Statistics Canada (1996). *Census of population (43 large urban centres, census tracts). Labour force activity, occupation and industry, place of work, mode of transportation to work, unpaid work.* Using E-STAT (distributor). estat.statcan.gc.ca/cgi-win/cnsmcgi.pgm.

Statistics Canada (2006). *Census of population (48 census metropolitan areas/census agglomerations and census tracts). 2006 cumulative profile.* Using E-STAT (distributor). estat.statcan.gc.ca/cgi-win/cnsmcgi.pgm.

Statistics Canada (2007). *2006 Census: Portrait of the Canadian population in 2006: Sub-provincial population dynamics.* http://www12.statcan.ca/census-recensement/2006/as-sa/97-550/p14-eng.cfm.

Stern, M. (2010). Inequality in the Internet age: a twenty-first century dilemma. *Sociological Inquiry, 80*(1), 28–33. http://dx.doi.org/10.1111/j.1475-682X.2009.00314.x.

Stevenson, D. (1999). Community views: Women and the politics of neighbourhood in an Australian suburb. *Journal of Sociology (Melbourne, Vic.), 35*(2), 213–227. http://dx.doi.org/10.1177/144078339903500206.

Stevenson, D. (2003). *Cities and urban cultures.* Maidenhead, Philadelphia: Open University.

Stevenson, D., Rowe, D., & McKay, K. (2010a). Convergence in British cultural policy: The social, the cultural, and the economic. *Journal of Arts*

Management, Law, and Society, 40(4), 248–265. http://dx.doi.org/10.1080/10632921.2010.500926.

Stevenson, D., McKay, K., & Rowe, D. (2010b). Tracing British cultural policy domains: Contexts, collaborations and constituencies. *International Journal of Cultural Policy, 16*(2), 159–172. http://dx.doi.org/10.1080/10286630902862646.

Stilgoe, J. (1988). *Borderland: Origins of the American suburb, 1820–1939.* New Haven, CT: Yale University Press.

Stolarick, K., & Florida, R. (2006). Creativity, connections, and innovation: A study of linkages in the Montreal region. *Environment and Planning A. Society and Space, 38*(10), 1799–1817. http://dx.doi.org/10.1068/a3874.

Strangleman, T. (2001). Networks, place, and identities in post-industrial mining communities. *International Journal of Urban and Regional Research, 25*(2), 253–267. http://dx.doi.org/10.1111/1468-2427.00310.

Strong-Boag, V., Dyck, I., England, K., & Johnson, L. (1999). What women's spaces? Women in Australian, British, Canadian and US suburbs. In R. Harris & P. Larkham (Eds.), *Changing suburbs: Suburban foundation, form and function* (pp. 168–201). London: E & FN Spon.

Sudjic, D. (1992). *The 100 mile city.* New York: Harcourt Brace.

Swyngedouw, E. (2007). The state of the situation. Post-political cities. In P. Stouthuysen & J. Pille (Eds.), *The state of the city: The city is the state* (pp. 203–227). Brussels: Brussels University Press.

Tavernise, S., & Gebeloff, R. (2010, 10 December). Immigrants make paths to suburbia, not cities. *New York Times.* www.nytimes.com/2010/12/15/us/15census.html.

Teaford, J. (1997). *Post-suburbia: Government and politics in the edge cities.* Baltimore: Johns Hopkins University Press.

Thomas, G. Scott (1998). *The united states of suburban: How the suburbs took control of America and what they plan to do with it.* Amherst, NY: Prometheus Books.

Thrift, N. (2006). Space. *Theory, Culture & Society, 23*(2–3), 139–146. http://dx.doi.org/10.1177/0263276406063780.

Titley, B. (1999). *The frontier world of Edgar Dewdney.* Vancouver: UBC Press.

Tonkiss, F. (2005). *Space, the city, and social theory: Social relations and urban forms.* Polity Press.

Törnquist, G. (1983). Creativity and the renewal of regional life. In Anne Buttimer (Ed.), *Creativity and context.* Lund: Lund Studies in Geography, B. Human Geography, No. 50.

Toronto Arts Foundation. (2009). *Best Buzz: An Executive Summary of an Exploration of Good Practices in Neighbourhood Arts Hubs in Toronto.* Toronto: Toronto Arts Foundation.

Tronrud, T.J. (1993). *Guardians of progress: Boosters and boosterism in Thunder Bay, 1870–1914*. Thunder Bay: Thunder Bay Historical Museum Society.

Tuan, Y.-F. (1974). *Topophilia: A study of environmental perception, attitudes, and values*. Englewood Cliffs, NJ: Prentice-Hall.

Turner, C. (2008, 9 February). Guerrilla barley growers go against the grain. *Globe and Mail*, p. F8.

Turok, I. (2005). Cities, competition, and competitiveness: Identifying new connections. In N. Buck, I. Gordon, A. Harding, and I. Turok (Eds.), *Changing cities: Rethinking urban competitiveness, cohesion, and governance* (pp. 25–43). New York: Palgrave Macmillan.

United Way of Greater Toronto. (2004). *Poverty by Postal Code: The Geography of Neighbourhood Poverty, 1981–2001*. Toronto: The Canadian Council on Social Development.

Urban Cultures Ltd. (1994). *Prospects and Planning Requirements for London's Creative Industries*. London: London Planning Advisory Committee.

Vachon, G., Luka, N., & Lacroix, D. (2004). Complexity and contradiction in the aging early postwar suburbs of Quebec City. In K. Stanilov & B.C. Scheer (Eds.), *Suburban form: An international perspective* (pp. 38–60). New York: Routledge.

van Heur, B. (2009). The clustering of creative networks: Between myth and reality. *Urban Studies, 46*(8), 1531–1552. http://dx.doi.org/10.1177/0042098009105503.

van Heur, B. (2010). *Creative networks and the city: Towards a cultural political economy of aesthetic production*. Bielefeld: Transcript.

van Heur, B., & Lorentzen, A. (Eds.). (2012). *Cultural political economy of small cities*. New York: Routledge.

Vaughan, L., Griffiths, S., Haklay, M., & Jones, C. (2009). Do the suburbs exist? Discovering complexity and specificity in suburban built form. *Transactions of the Institute of British Geographers, 34*(4), 475–488. http://dx.doi.org/10.1111/j.1475-5661.2009.00358.x.

Vaughan, R.M. (2006). Live without culture: An apology, on the occasion of a recent art project. In J. Dovercourt, C. Palassio, & A. Wilcox (Eds.), *The state of the arts: Living with culture in Toronto* (pp. 24–27). Toronto: Coach House Press.

Vaz, L.F., & Jacques, P.B. (2007). Contemporary urban spectacularisation. In J. Monclus & M. Guardia (Eds.), *Culture, urbanism and planning* (pp. 241–253). Aldershot: Ashgate.

Vicario, L., & Monje, M.M. (2003). Another "Guggenheim effect"? The generation of a potentially gentrifiable neighbourhood in Bilbao. *Urban Studies, 40*(12), 2383–2400. http://dx.doi.org/10.1080/0042098032000136129.

Vinodrai, T. (2006). Reproducing Toronto's design ecology: Career paths, intermediaries, and local labour markets. *Economic Geography, 82*(3), 237–263. http://dx.doi.org/10.1111/j.1944-8287.2006.tb00310.x.

Waitt, G., & Gibson, C. (2009). Creative small cities: Rethinking the creative economy in place. *Urban Studies, 46*(5-6), 1223–1246. http://dx.doi.org/10.1177/0042098009103862.

Walker, S. (2009). *Gordon Matta-Clark: Art, architecture, and the attack on modernism*. London: IB Tauris.

Walks, A. (2001). The social ecology of the post-Fordist/global city? Economic restructuring and socio-spatial polarization in the Toronto urban region. *Urban Studies, 38*(3), 407–447. http://dx.doi.org/10.1080/00420980120027438.

Wasserman, S., & Faust, K. (1994). *Social network analysis: Methods and applications*. Cambridge: Cambridge University Press. http://dx.doi.org/10.1017/CBO9780511815478.

Watson, A., Hoyler, M., & Mager, C. (2009). Spaces and networks of musical creativity in the city. *Geography Compass, 3*(2), 856–878. http://dx.doi.org/10.1111/j.1749-8198.2009.00222.x.

Watt, P. (2009). Living in an oasis: Middle-class disaffiliation and selective belonging in an English suburb. *Environment and Planning A, 41*(12), 2874–2892. http://dx.doi.org/10.1068/a41120.

Webster, R. (2000). Introduction: Suburbia inside and out. In R. Webster (Ed.), *Expanding suburbia: Reviewing suburban narratives* (pp. 1–13). New York: Berghahn Books.

Weisman, L.K. (1994). *Discrimination by design: A feminist critique of the man-made environment*. Urbana, Champaign: Illini Books, University of Illinois.

White, M. (2008). Condomanium! Condominium culture and cities of convenience. In P. Poletto, P. Beesley, & C. Molnar (Eds.), *Ourtopias: Cities and the role of design* (pp. 21–28). Toronto: Reverside Architectural Press.

Whitehand, J.W.R., & Carr, C.M.H. (2001). *Twentieth-century suburbs: A morphological approach*. London, New York: Routledge.

Whitzman, C. (2009). *Suburb, slum, urban village*. Vancouver: UBC Press.

Williams, R. (2004). *The anxious city: English urbanism in the late twentieth century*. London, New York: Routledge.

Wilson, D., & Keil, R. (2008). The real creative class. *Social and Cultural Geography, 9*(8), 841–847. http://dx.doi.org/10.1080/14649360802441473.

Wilson, E. (2000). *Bohemians: The glamorous outcasts*. New Brunswick, NJ: Rutgers University Press.

Winsa, P., & Rankin, J. (2012, 11 March). Patrol zone 121. *Toronto Star*, pp. IN1, IN3–5.

Wright, F.L. (1958). *The living city.* New York: Horizon Press.

Wright, H. (2008). *Instant cities.* London: Black Dog.

Young, D., Wood, P., & Keil, R. (Eds.) (2011). *In-between infrastructure: Urban connectivity in an age of vulnerability.* Toronto: Praxis (e)Press http://www.praxis-epress.org/IBT/inbetween.pdf.

Zakon, R. (1978). *The artist and the studio in the eighteenth and nineteenth centuries.* Cleveland: Cleveland Museum of Art.

Zukin, S. (1982). *Loft living: Culture and capital in urban change.* Baltimore: Johns Hopkins University Press.

Index

Note: Entries followed by f refer to figures. For example, 24f. Entries including n refer to notes. For example, 34n2, page 34, note 2.